# HOMO SOVIETICUS

*Gehirnprothesen* © 2013 Merve Verlag Berlin

This translation © 2017 Massachusetts Institute of Technology
All rights reserved. No part of this book may be reproduced
in any form by any electronic or mechanical means (including
photocopying, recording, or information storage and retrieval)
without permission in writing from the publisher.

This book was set in Helvetica Neue Pro by The MIT Press.

Library of Congress Cataloging-in-Publication Data

Names: Velminski, Wladimir, 1976- author.
Title: Homo sovieticus : brain waves, mind control, and telepathic destiny /
    Wladimir Velminski ; translated by Erik Butler.
Description: Cambridge, MA : MIT Press, [2017] | Includes bibliographical
    references and index.
Identifiers: LCCN 2016023014 | ISBN 9780262035699 (paperback : alk. paper)
Subjects: LCSH: Propaganda, Soviet. | Communism and psychology. | Communism
    and science. | Communism and society. | Soviet Union--Civilization. |
    Soviet Union--Social conditions.
Classification: LCC DK269.5 .V45 2017 | DDC 133.8/9--dc23 LC record available at
https://lccn.loc.gov/2016023014

For my parents.

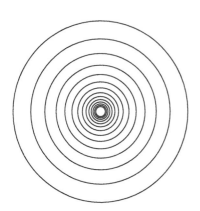

Translated by Erik Butler

Wladimir Velminski

# HOMO SOVIETICUS

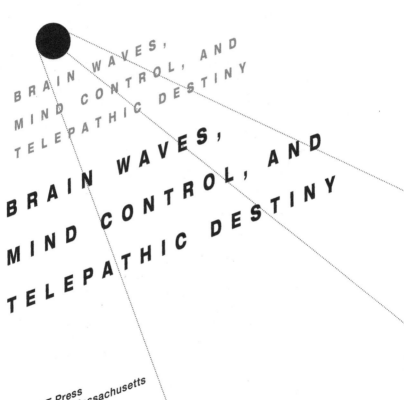

BRAIN WAVES,
MIND CONTROL, AND
TELEPATHIC DESTINY

The MIT Press
Cambridge, Massachusetts
London, England

# Contents

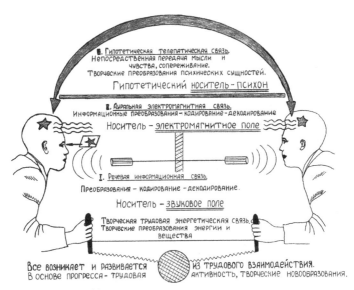

Material foundations of telepathy

# 1

## Preliminary Settings: Material Foundations

The question of how to educate and control the human
being, how to improve and perfect physical and
mental construction, poses an enormous problem that
can only be understood on the basis of socialism.
**Leon Trotsky**

The conscious and intelligent manipulation of
the organized habits and opinions of the masses
is an important element in democratic society.
**Edward Bernays**

Auxiliary supports for gray matter can entail the diminished ability
to think individually. A gesture meant to enlighten may give rise to
a mode of influence setting collective brainwashing into motion. In
figure 1.1, the stars where brains should be point to canny political calculations factoring into would-be modern blessings. This
ballpoint sketch, made by the Soviet cybernetic scientist Pavel
Gulyaev in 1965, shows a control circuit linking two figures. At the
lower edge of the picture, a saw cutting into a tree trunk checks
the loop; two identical-looking men are working the handles. A
red arrow arches over the figures; its points aim at their heads. In
each case, a star is shining where thought occurs. A Soviet star: a
neural prosthesis.

The drawing's subtitle reads *Material Foundations of Telepathy*.
Inside the control circuit, Gulyaev lists cybernetic field systems. The
first, *rhetorical-informational switching* in the auditory sphere, is
a matter of reorganization, coding and decoding; the drawing
presents the circuitry as a sender-receiver network. The second is
*auratic-electromagnetic switching*, which also concerns information-
processing, coding and decoding; its medium, the drawing shows,
is an electromagnetic field.

The waves emanating from inside the figures' heads make the iconic significance of the circuitry plain: having received a visual transmission (depicted in a rhombus containing a star), the figure on the left is coding this message as electromagnetic impulses and broadcasting it again; to the right, the other figure—the receiver—is decrypting it. Knowledge is being conveyed via thought based on material transmission technologies; it is directly registered on the receiving end. The stars where brains should be indicate that mental transfer has been politically instrumentalized through and through: the scene legitimates censorship and control on the basis of established scientific insight and the speculation of research.

The design indicates that when *rhetorical-informational* and *auratic-electromagnetic* systems are perfectly guided, the third attribute of *Homo sovieticus* emerges: a *hypothetical-telepathic circuit*. Represented by a straight line, its functionality comprises the unmediated transmission of thoughts, feelings, and sensations; at the same time, it involves the creative reorganization of psychic substance. *Psikhon* is Gulyaev's word for the energy coursing through this circuitry; clearly, its properties take shape in a *feedback* system that alternates between *control over* media and *being controlled by* media.

The following discussion of neural prostheses does not concern twenty-first-century medical implant methods. Nor does it concern electrodes or brain pacemakers set up to stimulate thoughts.[1] Instead, the focus falls on processes that act on people from outside. That said, it is impossible to draw a dividing line between the two realms inasmuch as the practices described are directly connected to the research practices of the era and were in fact implemented within such a framework to enhance creativity, improve memorization, and heighten performance. In this context, the use of cerebral prostheses did not involve individualized courses of treatment to promote original thinking. Instead, the messages coded as stars were meant to reach the brain of a whole society, propagate there, and establish uniformity of thought

[*Gleichschaltung des Denkens*]. As such, the term *neural prostheses* does not refer to technology that brings about an aura of the mechanical so much as actual designs that effect switching, steering, and control.

The dialectic at work connects with the idea of a "prosthetic God" that Sigmund Freud discusses in *Civilization and Its Discontents*.[2] For all that, the study at hand does not address the "natural aspirations of culture"; instead, it explores how these aspirations were influenced and guided.[3] Although the psychic apparatus of the new Soviet collective was a construct, the operations of the "soul" were determined and directed by electromagnetic stimuli more than by dreams.[4] "Communism is Soviet power plus the electrification of the whole land," Lenin's slogan declared. These words were to become the motto for the people as a whole.

Gulyaev's schematic follows the psychobiological vision of human perfection that Leon Trotsky—Lenin's closest confidant and intended successor—announced to intellectuals and artists in 1923:

> Man at last will begin to harmonize himself in earnest. He will make it his business to achieve beauty by giving the movement of his own limbs the utmost precision, purposefulness, and economy in his work, his walk, and his play. He will try to master first the semiconscious and then the subconscious processes in his own organism, such as breathing, the circulation of the blood, digestion, and reproduction; within necessary limits, he will try to subordinate them to the control of reason and will. Even purely physiological life will become subject to collective experiments. The human species, the coagulated *Homo sapiens*, will once more enter into a state of radical transformation, and, in his own hands, become an object of the most complicated methods of artificial selection and psychophysical training.[5]

Until Trotsky was declared the public enemy of the Soviet Union, he devised projects that ran counter to the plans of the Central Committee (which Stalin dominated). Accordingly, the New Man was to "make it his purpose to master his own feelings, to raise his instincts to the heights of consciousness, to make them transparent, [and] to extend the wires of his will into hidden recesses" of his being.[6]

Thirty years later, Gulyaev built on this very program to create a higher social-biological type, integrating futurist and constructivist ideas of the 1920s into a state-of-the-art cybernetic mechanism. The iconography of his drawing points to psychological and physiological experiments on human subjects at the borderlands of science: a control system located simultaneously in the political and social spheres. As depicted, this system focuses on an interconnected and optimally calibrated form of polarity: two citizens' combined and coordinated efforts to saw through a log blocking their way.[7] Absolutely synchronized equality results from this relationship, and a new medium takes form. Hereby, the function of labor plays a central role. Even if work and associated ergonomic operations have not been integrated into the circuitry, they still provide the foundation as creative-dynamic-energetic connectivity. Work stands for the productive transformation of energy and matter. Everything comes into being and develops through reciprocal labor, the text just under the saw affirms. Work to create the new constitutes the basis of the constructive process.

These same, material foundations for an immaterial medium stand at the center of the pages to follow. Gulyaev's writings, held at the Academy of Sciences in St. Petersburg, serve as the compass throughout.[8] The point of departure is provided by the ergonomic operations outlined by labor scientist Aleksei Gastev (1882–1939), which were repurposed as creative-dynamic-energetic circuits in the language-oriented, incorporeal practices that the neurophysiologist Vladimir Bekhterev (1857–1927) implemented in a laboratory setting to engineer rhetorical-informational circuits.

In turn, the patents and designs of television pioneer Hovannes Abramovich Adamian (1879–1932) represent *auratic-electromagnetic networks*. Linked directly to Bekhterev's experiments, they provide the basis for *hypothetical-telepathic circuitry*.

In this light, the psycho-hygienic prescriptions and cybernetic wiring in Gulyaev's design amount to a technological process for exercising influence and control. The underlying constellation of forces comes into view through the mediating circuitry that steers operations; through this lens, it is possible to describe and analyze the emergence of immanent strategies of power, apparatuses for influencing, methods of surveillance, and paranoid modes of thought. The point is not to identify yet another "science of man" during the Soviet era—much less to practice such a science. Rather, the goal is to outline a new picture of the thinking that shaped the age.

Such an approach—which ties the construction of epistemological objects to the ways they are presented to, and staged for, the public—will enable us to discern creative dispositives beyond the confines of the laboratory. Inasmuch as the truth claims of these paraphysical sciences offer aesthetic alternatives—as in a work of fiction—our attention turns, again and again, to the political and metaphorical contaminations that occur in the process of transfer.[9] In other words, the designs examined in the following did not just function reactively, as means for conveying simplified versions of difficult scientific insights; rather, they represented ways that new findings were situated—and transformed—in a process of illustration, decontextualization, and recontextualization.[10] Especially in the Soviet Union, such plans did not unfold in a politically neutral space. Instead, they served to spread and instrumentalize ideological positions and aesthetic precepts, with contagious epistemological effects.

On the one hand, then, the coupling of science and poetology points back to human beings who, in receiving an aesthetic transmission, are meant to understand the neurological basis of the

*New Man* and make it their own. On the other hand, this same feedback loop affirms the postulates of a controversial science whose foundations are a matter of political calculation.

By way of historical thought-experiments—which do not belong to a metaphorical register in this context—our study explores how the phantasms haunting scientific research were enlisted to steer thinking and manipulate the population. The heyday of such practices of suggestion occurred when the regulatory cycle of the Soviet Union was threatening to fail: popular elements, already suspicious of the system, needed to undergo treatment in order to be "recharged."

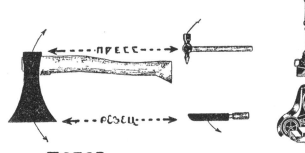

**2.1**

Innovation processes according to Gastev
"The axe…
    this is the eternally young grandfather
        of the jackhammer
            the hydraulic press
                the metalworking lathe
                    the grinding machine"

# 2

■ **Ergonomic Operations**

This thunder will either scare people to death,
make them collapse, or they will grow into
something greater.

**Aleksei K. Gastev**

## 2.1 The Ornament of Regularity

For Gulyaev, *creative-dynamic-energetic* switching provided the
foundation for bipolar control circuitry. With this designation, the
cybernetic scientist was referring to the work of an intellectual
predecessor who had fallen victim to Stalinist cleansing: Aleksei
Kapitonovich Gastev, the encomiast of futurism and propagan-
dist of Taylorism in the USSR. In one of his manifestos, from 1924,
Gastev had proclaimed that the inventor should be a dreamer. He
did not have in mind a gentle and idyllic romantic, "fantasizing
about rivers of milk and the Land of Cockaigne." Rather he envisioned an "austere, flexible, and constructively thinking dreamer
who can quickly discern the relationship between one phenomenon
and another."[1]

Gastev's enjoinder to creativity, programmatically entitled *How
to Invent* (*Kak izobretat'*), underscores the necessity of calibrating work on all levels; more still, it outlines the qualities required
of an inventor. The latter should be able to "cycle through, in a
moment, all conceivable variants and remember, as quick as light-
ning, whether the same thing has already occurred on another

machine or device, in another instance."[2] This statement makes it clear that Gastev did not believe, as his contemporary Kazimir Malevich held, that it was a matter of investing the "power" of "any given, creative human being" in the "discovery of a method to overcome our endless progress (*preodoleniya beskonechnogo prodvizheniya*)"—that is, to put an end to further development, invention, labor, and creation.[3] Quite the contrary: Gastev sought to promote new inventions by reflecting on mechanisms already in place, which would ensure further advances (figure 2.1). The *creative-dynamic* message broadcast by the founder of the Central Institute of Labor (*Tsentral'nyi institut truda*, *TsIT*)—which diffused the doctrine of biomechanics and the system of Taylorism in factories, theaters, and athletics[4]—bore the traits of its author. In offering visions of creativity at work, which comprises the art of invention and production,[5] Gastev displayed his twofold nature. As a poet, he participated in the futurist project of engineering a new machine-man; at the same time, he was the agent of capitalist Taylorism—that is, of tough and hierarchical ergometry.[6]

The October Revolution inspired Gastev to reconfigure all of the disciplines relevant to the nascent regime.[7] His proximity to Lenin permitted him to implement his ideas, too. The *scientific organization of labor* (*nauchnaya organizatsiya truda*, *NOT*) that he developed reached deep into realms close to those claimed by the Russian avant-garde—especially its radical, constructivist variant. At the same time, it connected with the government's New Economic Policy.[8] Accordingly, the following considers the interference ratio between Gastev's visions for labor and the futurist-proletarian messages he also broadcast. Our focus is the oscillation between these two poetological configurations, from which the symbolic, precybernetic gestalt of Gastev's integrated program emerged. Here, thirty years later, Gulyaev would install the *creative-dynamic-energetic circuitry* he had developed experimentally.

## 2.2 The Construction of Thought

Aleksei Gastev—by turns, worker, activist, poet, revolutionary, scientist, visionary, and public enemy—was born in Suzdal in 1882. By the age of eighteen, he was already gaining experiences as a day-laborer in Paris, Berlin, St. Petersburg, and Kharkov.[9] Then, in a time of transformation and upheaval, he was reborn among factory machines as a poet and visionary. "We Grow Out of Iron" (*My rastem iz zheleza* [1914]) represents his first exploration of the theme:

⑪

> Look! I stand among workbenches,
>
> hammers, furnaces, forges, and among a hundred comrades,
> Overhead hammered iron space.
>
> Fresh iron blood pours into my veins.
> I have grown taller.
> I too am growing shoulders of steel and arms immeasurably strong.
> I am one with the building's iron.[10]

"Growth" and "standing tall" stemmed from Gastev's work at the Siemens & Haske Telegraph Construction Company in St. Petersburg.[11] The machinery—"its enormous frames displaying the plant's strength in well-equipped assembly halls"[12]—defined his "setting" in relation to poetry.[13] Gastev's next setting came from public transportation: monitoring streetcars and checking people's tickets. Here, his lyricism took the form of left-wing expressionism.

After the October Revolution, Gastev embraced the *Proletkult* movement.[14] Events in 1917 gave rise to complex, interdisciplinary conditions favoring the institutional exchange of scientific practices and methods.[15] It was not the United States or Western Europe where researchers proclaimed a *cultural* mission, but the land where the first *Arbeitsstaat* (Ernst Jünger) was in the course of emerging. At the Central Institute of Labor in Moscow, laboratory technicians investigated human perception in terms

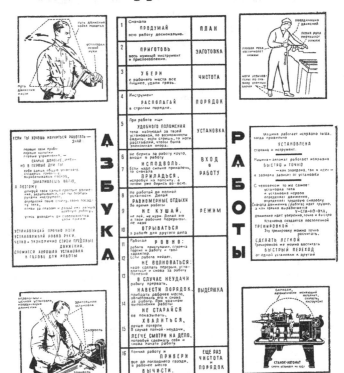

**2.2**

The alphabet of labor

of physiological and psychological criteria.[16] After the institute's founding, which he deemed his "final artistic production," Gastev wrote only journalistic and scientific works.[17]

Gastev found the principles for his vision of rational and thoroughgoing organization in Frederick Winslow Taylor's system of labor efficiency.[18] Individualized job sequences, shift-planning, calibrating workers' motor skills to the operations of machinery, and a rhythmical order in scheduling arrangements constitute the principles of Taylorism. A purely scientific approach is meant to lead labor, management, and the enterprise as a whole to function optimally, thereby ensuring "prosperity for all." The same "settings" underlie Gastev's visions: thorough organization of factory space, measuring tasks and outcomes, and, above all, (re)training workers. Regulated sequences of work and movement—more specifically, sequences that have been *thought out in advance*—are meant not just to maximize performance, but also to engineer the fusion of man and machine.

The first articulation of "training-agitation"—an "alphabet of labor" eventually displayed at workplaces throughout the Soviet Union (figure 2.2)—already codified the importance Gastev attached to orderly thought processes. "Before getting to work, one must first think [the operation] through—think it through so that a model of the finished task and the complete sequence of steps are present to mind."[19] That said, the fusion of man and machine also meant turning away from Taylorism. *Setting* the labor process does not mean competition to be the best—which was Taylor's principle: money motivates, and pay depends on performance. Instead, the point is for each worker to get the best out of himself (or herself), across the entire system.[20]

Needless to say, social status, familial belonging, and personal relations play no role when it comes to employing technological orders of experimentation in order to realize individual aptitude.[21] Gastev justified his laboratory arrangements and research plans by understanding labor as a *form* susceptible to theoretical and practical intervention. He sought to exercise "active influence on the human being":

> By creating determinate sequences of settings, reconstructing them, and improving them—always in relation to laboratory experiments and the experience of production—we will advance toward exercising active influence on the organism, training it, and creating new, organized reflexes.[22]

The Soviet labor scientist viewed the human being as a program; not only movements but also thought-processes were to be optimized. Practical experience had taught him that this doctrine "meets with certain insurmountable conditions that qualify as inborn and do not admit a radical change." "Such dynamic sluggishness," he suggested, should be counteracted by "refitting" that would enable workers to "rally for a further, radical offensive (*nastupleniye*)":

> Then we will exercise influence on the organism so that it does not primarily occur in the sphere of movement. Rather, we will have to ask the question of biological, surgical impact on the organism—whether this involves active manipulation of the metabolism, the attempt to change the elements of the blood, the effort to alter circulation, or influencing the spheres of secretion, symptoms, or nerves. From this vantage point, we can affirm that our doctrine seeks heightened activity in the realm of the biological sciences.[23]

The economy and efficiency of machines are to produce an effect on the new, virtuous consciousness of humankind. The goal is selfless devotion to work and the collective, to the point where processes merge with the laboring masses: "The worker has organically fused with the factory mechanism as a whole."[24] The factory is a temple, the machine a new God, and Gastev the Deity's singer and herald.

As it turns out, this same, hybrid arrangement of projection and programming would prove fatal for the constructor of the New Man.[25]

## 2.3 A New Cultural Setting

Until then—so long as Gastev could still pursue the scientific organization of labor—the concept of *setting* (*ustanovka*) was key. He picked the term in the capacity of a "practitioner and technician":

> The word *setting* has already been in use for some time in technical language. Initially it was known only to mechanics who worked on calibrating machines and electrical equipment. ... It involved, first, setting up the framework for a given machine, erecting all the parts necessary to make it operate, then adjusting the machine. But in recent times, in the last fifteen years, the word has left the world of assemblymen and has come to be used in factory production. A new group of workers has emerged who perform an administrative function, representing talented organizers of work; they are called "setting engineers" (*ustanovshchiki*).[26]

(15)

**2.3**
Strategies of observation

Gastev stresses that the matter concerns both the concrete, variable *set-up* of the workbench and the *bearing* of the worker using the machine, who adjusts himself to its mode of functioning (figure 2.3, detail of figure 2.2). At the same time, he explicitly invokes a *new cultural disposition* (*novaya kul'turnaya ustanovka*); thought-processes must be calibrated with operative settings in order to heighten the productivity of the mind:

16

> Hammer, tongs, wheel, pencil, matches, staves of wood—all of these things must be explored with a keen eye for the sensation they offer, of which the layman (*obyvatel'*) suspects nothing. We must create "makers" of *culture*, not the authors of those beloved compilations of ideas now flooding the shops, but rather talented creators who mount practical systems in all cultural spheres. It is a matter of love for labor, for the constructive ease of physical toil, for cultivated artistic work. The task is to turn things around. It concerns a sphere of culture that has never existed before.[27]

In this sense, Gastev sought to bring about a *comprehensive setting*, by which means "any youth can quickly learn any profession at all."[28] This objective represented a further point of difference with the Taylorist model. The latter, Gastev observed, does not attach enough importance to learning or integrating basic motor operations into the labor process; Taylorism is overly enamored of virtuosic displays.[29]

Gastev, in contrast, knew that Man's new setting—giving one's best—can only function if one believes in something and "loves" it. Selfless devotion to labor presupposes aestheticization and popularization, then. He voiced this goal in poetic form as a machine-romance, "I Learned to Love" (*Ya uchilsya lyubit'*, 1917):

> I learned to love you, iron din,
> The festive ringing of steel and stone.
> Lava ... unquiet, restless fire,
> The hymns of machines, their splendid tone.[30]

The merging of love with scientifically tested and scientifically rehearsed cultural practices was meant to reveal a host of new outlooks: settings for work discipline and creativity without precedent.

The parallels Gastev drew to the life sciences also command attention:

> In modern biology, the term *setting* is being employed more ⑰ and more frequently. Even though its use has not yet been generally established, still it can be said that the psychological term *setting* has been circulating for some time now in the sciences. In German schools of psychology, the term *setting* is everywhere.[31]

Gastev understood *setting* in the technical sense when conducting psychological experiments, too; here, Karl Krall's studies provided a point of reference.[32] He also enlisted Ivan Pavlov for support, noting that "the concept of the unconditioned reflex can be understood as a natural, inherited structure."[33] It followed that a setting "resulting from a known practice"[34] qualifies as a conditioned reflex.

In drawing this analogy between Pavlov's doctrine and Krall's studies of animal training, which he connected in terms of reflexology, Gastev sought to defend both researchers against the charge of mere rote training.[35] For him, their research proved that regulating conditioned reflexes would yield methods for optimizing labor processes. Accordingly, "setting" at the *TsIT* also signified "the combination, in the 'living machine,' of determinate unconditioned and conditioned reflexes into a unified complex of 'working production' indispensable for the successful execution of operations in any profession at all."[36] Alertness, manual skill, willpower, strict order, organization, and imaginativeness were to provide its foundations.

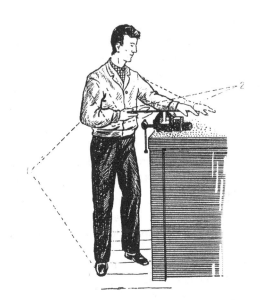

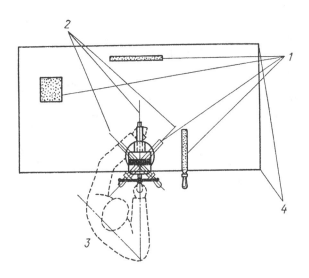

## 2.4 Strategies of Self-Observation

Gastev attached particular value to *self-observation*, both in a technical sense—at the workplace—and in broader terms of culture (figures 2.4 and 2.5):

> In order for a good worker to observe his operations, special training is necessary. ... Truth be told, it is very difficult to work without self-observation. All people who have ever learned a skill in earnest had to master work in advance and practice self-observation.[37]

(19)

In conjunction with the categories of setting Gastev elaborated, this second-order strategy of observation gave rise to a new way of seeing, coextensive with the concerns of cybernetics. Still before Heinz von Foerster devised a new theory of how the operations of the central nervous system produce subjective reality—and came to speak of the *observation of observation*—Gastev had worked out a mode of visual engagement for ordering intellectual movement in progressive fashion (figure 2.5). Observation that is simultaneously generated internally and produced externally creates a feedback loop: dynamic self-perception. Establishing an outside point of observation means setting up a visual template for registering one's own activities. This "paranoid" alignment is meant to generate constant counterinsurance—after all, another orientation-template is already at work in mental operations. Not only will it govern the realms of technical labor and culture; it will also bear on human evolution.

**2.4**
Static and kinematic work

**2.5**
Visual settings

## ТЕОРЕТИЧЕСКАЯ СХЕМА ОПЕРАЦИОННЫХ УСТАНОВОК

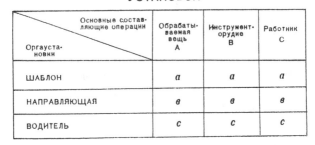

| Основные составляющие операции / Оргустановки | Обрабатываемая вещь A | Инструмент-орудие B | Работник C |
|---|---|---|---|
| ШАБЛОН | *a* | *a* | *a* |
| НАПРАВЛЯЮЩАЯ | *в* | *в* | *в* |
| ВОДИТЕЛЬ | *с* | *с* | *с* |

## С Х Е М А
### БИОЛОГИЧЕСКИХ УСТАНОВОК РАБОТНИКА

| АКТИВАТОРНЫЕ УСТАНОВКИ | СТАТИЧЕСКИЕ УСТАНОВКИ | ДИНАМИЧЕСКИЕ УСТАНОВКИ | СЕНСОРНЫЕ УСТАНОВКИ |
|---|---|---|---|
| Энергетический баланс работника | Корпусно-механическая установка | Двигательная культура работника | Зрительные, слуховые, осязательные реакции, конструктивно-мыслительные процессы |

**2.6**

Theoretical scheme of
operational settings

**2.7**

Scheme of the worker's
biological settings

In this context, Gastev speaks of changes in the organism occurring through constant activity and exercise:

> The whole history of Man (*Homo sapiens*) is a history of developing biological adaptation [*sozdaniye bioprisposobleniya*]. All the newest biological doctrines (Darwinism, conditioned reflexes, the doctrine of rejuvenation) have, strictly speaking, either studied the ensemble of biological adaptation (Darwinism) or created other methods of biological adaptation.[38]

Accordingly—and how could it be otherwise?—such adjustments represent further operational-automatized settings; they are templates for guiding and steering (figure 2.6).

Gastev stresses that their automatized quality should only "be understood in the biological, or better, neuromechanical sense of the word."[39] But whatever the vocabulary employed, workers' static and kinetic faculties admit tuning and modification. The *scheme of biological settings* is subdivided into four parts (figure 2.7). The first concerns activation, the initial energy required for production; the second, static setting involves material arrangements—that is, what remains immobile during work; in turn, a dynamic setting concerns the direction of mechanic-muscular operations; finally, there is a sensory setting for workers' visual, acoustic, tactile, and neural reflexes.[40] As the diagram illustrates, the coordinated steering of objects and knowledge, in a feedback loop of schematic self-observation, yielded a model (figure 2.8) destined for further expansion in the cybernetic age proper.

Gastev recognized the key difficulty his project faced[41]: a good worker "commands the automatism of movements" so fully that he is no longer capable of self-observation; that is, he stands "wholly under the power of automated nerves."[42] Ideally, however, the entire work process should amount to a kind of self-regulated control loop (figure 2.9). Accordingly, the human capacity for self-observation demands special attention.

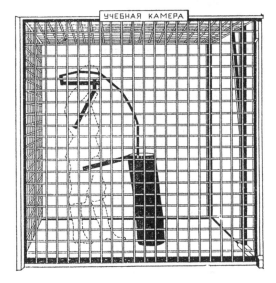

**2.8**
Labor grid

**2.9**
Learning chamber

Gastev's *Analysis of Operations* (*Analitika priemov*) explores various methods and techniques (e.g., static and dynamic mechanics, cinema, and photography) that bear on the autoregulation of motion sequences.[43] One approach is to follow an operation, intermittently slowing it down to obtain an integral picture of *completed motions* (figure 2.10). For instance:

㉓

If the gripping operation irritates the worker, this means that too much exertion is required of him. But if it irritates him minimally or provides a rhythm allowing certain muscles to relax, this offers a vast field of intervention for rationalizing motion. Our provision for hammering with a swing—more precisely, a mode of gripping that reduces the work of fingers struggling against centrifugal force—represents a clear instance of a proper setting.[44]

It is no accident that the terms *setting* and *operation* were also key terms in the literary theory of the avant-garde.[45] Precisely here, Gastev's visionary bearing toward invention and its connection to cybernetics comes into view.[46] With new biophysiological techniques and minutely controlled effects, the mind's efficiency is meant to undergo dramatic augmentation. The model of the striking implement—a hammer—is supposed to *set* new possibilities for thinking in place.

Just as an expertly used hammer can drive in a nail with a single blow, Gastev's visions were meant to deliver a dynamic charge to the new Soviet culture. His *Program for Cultural Settings* (*Programma kulturnoy ustanovki*) indicates that "observation automatically gives rise to the need for an exact art of representation."[47] From the outset, Gastev underscores the significance of language: "words must be brief, precise, and categorical. … The ability to report—as opposed to engaging in refined discussion—should stand as the armament (*vooruzheniye*) for those setting the course of the new culture."[48]

(24)

As a poet, Gastev devoted particular attention to this "hard language"—rhetoric derived from industry and the transportation system.[49] The "technical instructions" accompanying his *Pack of Orders* (*Pachka orderov*, 1921), also entitled *The Word in the Press* (*Slovo pod pressom*), declare:

> *Pack of Orders* is to be read in measured segments, as if they had been fed into a machine.
>
> In the reading there should be no expressiveness, pathos, or pseudo-classical verve; nor should any passages be pathetically intoned. Solemn emphasis.
>
> Words and phrases follow one another at the same tempo.
>
> A massive action is underway: the *Pack* offers the listener a libretto for concrete events.[50]

In the new cultural setting, the proper bearing toward *production and execution* meant extending artistic and aesthetic forms so that poetic language would ultimately prove indistinguishable from technical language. In other words, Gastev sought a linguistic strategy to convey vivid immediacy; it was exemplified by the innovation and expressiveness his own invention displayed.

## 2.5 The Setting for Invention

The Patent Office in Russia still holds documents outlining Gastev's vision. The inventor did not register a patent until 25 March 1925. A quick look at the history of the institution readily accounts for the delay. In post-czarist Russia, in the course of revolution and civil war, the system of *privilege* that had formerly governed intellectual property was dissolved. On 30 June 1919, Lenin signed a decree, *On Inventions*, which he had authored himself. It outlined four fundamental principles:

1. Introducing a new form of legal protection for inventions. This includes certifying authorship, which is meant to ensure optimal agreement between the private rights of the inventor and state interests.

2. Authorizing functions for government bodies connected with making inventions useful according to plan.

3. Affording conditions favoring the broad development of inventive creation and achieving, as fully as possible, results for everyday life with the goal of advancing the political economy of the nation.

4. Guaranteeing rights for the manifest creators of inventions, as well as stimulating them morally and materially.[51]

The question whether the decree led to the desired mental productivity may be left open.[52] At any rate, this foundation for the Soviet patent system set aside practices that had held until this point, and it declared that the state alone would derive profit. Only when the decree was modified at the end of 1924, in order to secure patent rights, did Gastev register his invention (even though he stood close to Lenin). The documents he filed include a *Description of a Device for Exercising the Joint of the Elbow or Wrist in Teaching Work with a Hammer*:

The present invention is intended to train the hand and elbow during work with a hammer and is meant to habituate the hand to the proper position. Figures 1 and 2 [figure 2.11] represent the proposed device in perspective. ... The function of the device concerns fixing the required hand position when laboring.[53]

26 Gastev's contrivance for optimizing hammer use, with its seemingly prosthetic quality, implies that the New Man of contemporary discourse should not be viewed as an *ideal.* Instead, as Sven Spieker has observed in a related context, "the orthopedic apparatuses of the Russian constructivists [mark] the point of intersection between the integral body and the deficient, incomplete body, which, as castrated or potentially castrated, requires symbolic completion again and again."[54] That said, Gastev's invention was not a common prosthesis, nor did it involve only the body. It served as a template for *setting* the mind. In other words, the device is a neuroprosthesis:

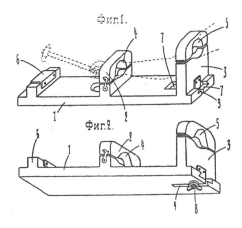

**2.11**
Sketch of the model
patented

The double, or split, perspective at the intersection between motor operations and machinery—mediated by the neuro-physiological switchboard of the brain (which is not visible here)—has retroactive effects on tasks that the laboring hand performs. External observation, in the sense of self-discipline, becomes integrated into self-observation.[55]

By means of Gastev's *device*, the worker is supposed to be calibrated in such a way that he operates as part of an organic system without outside intervention. Gastev's inventions did not aim to make workers into automata of muscles and nerves, merely drilled to perform natural operations. Rather, they were meant to synchronize processes of steering and regulation with all attendant settings (activation, static and dynamic mechanics, sensory circuits) so the ensemble would bear the imprint of well-defined regularity. As a corollary, new modes of thinking would arise from the template now in place.

From the outset, Gastev foresaw a new dynamic for regulating production; more still, his designs anticipated a shift to self-regulation through technical feedback loops, on the one hand, and neural processes, on the other. In this way, "a new generation of setting engineers for machines, transportation, assembly lines, and, finally, human beings"[56] would emerge. Needless to say, the point was not merely to discipline bodies; rather, in Michel Foucault's words, provisions would serve to regulate the population in its entirety: "The normalizing society is a society in which the norm of discipline and the norm of regulation intersect along an orthogonal articulation."[57] The specifications of Gastev's scientific organization of labor—no matter whether it involved writing at a desk, metalworking, or plowing a field—combined in his theory of the art of invention. *How to Invent* declares: "Whoever speaks seriously about the scientific organization of labor should know that one must necessarily be an inventor to carry it out." And inventing means establishing rules: "when you add a further rule, this means that you have integrated yourself into work."[58]

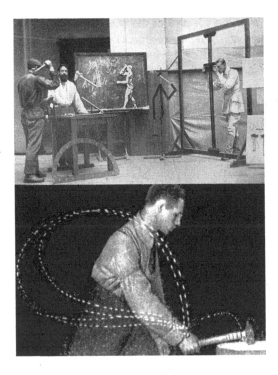

**2.12**
Positions and motions of work—
Gastev delivering a hammer blow

In sum: Gastev's notion of *setting* did not just fit workers' movements to the rhythm of machines, resulting in elegance and efficiency. The symbolic significance of his inventions also meant a new *setting* for viewing Meyerhold's experimental theater, Eisenstein's cinema, and contemporary literature. The inventor, Gastev maintained, must "have a great memory at his disposal, a pictorial memory"; more still, he needs a "lively capacity for imagination":

> The inventor must be able, after taking things apart in analysis, to set one phenomenon in relation to another, quickly and as fast as lightning, through the power of fantasy—to recognize similarities so that a flash of inspiration occurs.[59]

Gastev's "lively capacity for imagination," which "set[s] one phenomenon in relation to another," represents a strategy for intuitive clarity that does not focus on static images, but on sequences of images. This procedure was implemented practically by cinematographic means and employed in the laboratories of the *TsIT* for analytic purposes (figure 2.12). Tellingly, Gastev does not speak of pictures, but of phenomena that are linked to one another. The sequence of phenomena creates an atmosphere—then, "a moment of inspiration" strikes:

> Inspiration seizes [the inventor] just as he takes something apart, observes it precisely, and then, in an instant, with the aid of his great memory, quickly finds a similar phenomenon in other constellations and sets this phenomenon in relation to that one.[60]

The semantic range of *setting* and the art of invention it should bring about make it clear that Gastev was a pioneer of cybernetics and attendant epistemological orders. Instead of viewing labor in moral and physical terms—as effort spent and energy expended—he made it the object of experimentation and construction, with the goal that workplace operations would proceed by recoding and self-regulation. In this vision, labor, as *creative-dynamic-energetic switching*, constituted the basis for a system of self-monitoring circuitry. Labor stands for the *creative* transformation of energy and matter, because *everything* develops from interactive dynamism, with labor and creative generation forming the basis for the constructive process.

From the 1960s onward, Gastev's vision for constructive-creative processes served as a pre-cybernetic model in Russia. It provided the foundation for new control practices (as depicted in Gulyaev's illustration, figure 1.1) and, as such, factored into the design of the first computer work stations in the USSR. But during the Stalin era, the same designs led to his undoing. On 8 November 1938, Aleksei Gastev was arrested for "counterrevolutionary terrorist activities." On 14 April 1939, the death sentence was pronounced. The next day, in a Moscow suburb, the inventor was shot.

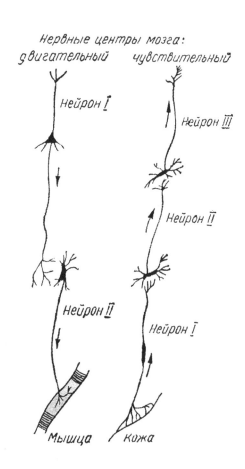

**3.1**

Ganglia according to Bekhterev.
"Nerve centers of the brain:
Moving (muscle)—feeling (skin)"

# 3

## ▬ Immaterial Practices

The *contagium vivum* of bodily infections, whose
nature and effects research is making clearer
and clearer, has its counterpart in a *contagium
psychicum*; even if it escapes all observation
by rude sensory means, the latter, as much as the
former, still threatens the human organism with
the danger of immediate infection.

**Vladimir Bekhterev**

### 3.1 The Diagnosis of Power

The history of immaterial practices cannot be detached from leg-
endary *stories* about them; they lend the controversial science
both power and pathos. One of the *countless* stories concerns
the neurologist and psychiatrist Vladimir Mikhailovich Bekhterev,
who was directly involved in telepathy research. On 23 December
1927, Bekhterev showed up late to a medical congress in Moscow
about the *Pantheon of Brains* (*Panteon mozga*), which he had initi-
ated. He justified his tardy arrival by declaring, "I was examining a
rough-handed paranoiac."[1] After the panels, the delegates decided
to visit the theater and see a play. During the intermission, two
unknown men approached Bekhterev and ate with him at the buf-
fet. After the production, Bekhterev took leave from his colleagues
and went to his hotel. He never woke up.

Family members, friends, and colleagues could only guess who
the "rough-handed paranoic" was—or the reason for Bekhterev's
death. It is said, however, that the scientist had received a telegram
from the Medical Division of the Kremlin shortly before leaving
Leningrad for Moscow; it requested that he announce his arrival in

the capital.[2] What is more, people knew that the All-Union Communist Party of Bolsheviks had commissioned Bekhterev to draw up a psychological profile of the general secretary of the Central Committee; indeed, the doctor had already made inquiries among Stalin's relatives and confidants. Witnesses attest that Bekhterev delivered his diagnosis—*paranoia*—on 23 December 1927, the same day he encountered the mystery men. Whatever happened, among the first brains displayed in the *Pantheon* was Lenin's—along with Bekhterev's own.[3]

Detective work to determine the accuracy of this tale may be left to others. Our attention will fall on rhetorical and informational traces of the Bekhterev *case.* One year after the scientist's mysterious death, the results of his research entered the literary realm. During this same period, other investigators conducted experiments on telepathy and came to view the brain as a broadcasting tower able to emit, receive, and, ultimately, influence thoughts.[4] The precise status of telepathy in science does not stand at issue, nor do specific differences between research efforts. It is enough to note that practices displayed analogies and points of overlap under these historical circumstances.

In 1929, Aleksandr Romanovich Belyaev—a founding figure of Soviet science fiction—published *The Ruler of the World* (*Vlastelin mira*). This novel not only represented illusions of power tied to telepathy, it also transferred the technological *episteme* of the age to readers. When orders of experimentation were connected to a literary circuit, complementary resonances ensued, whose effects included further real-world experimentation.

In other words, it is not just a matter of what Friedrich Kittler demonstrated in *Discourse Networks*: that literature was losing its monopoly over the generation, communication, and recording of secrets of the soul to the graphic operations of psychophysiological apparatuses,[5] but also that in this case, telepathy provided the method for effecting such a transfer. Besides practices and processes occurring in sealed-off laboratories, literary texts in the public sphere occasioned an array of interactions that defined the shape and significance of immaterial thought-transmission.

## 3.2 Rulers and Their Subjects

> I call Lord [a dog]; he comes to me. I take his head in my hands, as if to emphasize, symbolically, that he stands utterly under my power: he must repress his will entirely and become an automaton, the unjudging executor of my will.[6]

(35)

So begins the description of how the famous animal trainer Vladimir Durov taught an animal to run to a table, take a book in its teeth, and bring it back to him. In 1919, at a conference hosted by the Institute for Brain Research on "'Mental Influence on the Behavior of Animals," Bekhterev presented Durov's meticulous procedures to his colleagues.

Following a survey of national and international findings (and after charging the Russian hypnotist Dr. Naum Kotik with charlatanry[7]), the speaker described how he had met the trainer; then he detailed the experiments he and associates conducted with him. Bekhterev's account of Durov's method continues:

> To achieve this, I fasten a severe gaze on [the dog], as if its eyes had fused with my own. The dog's will is paralyzed. I concentrate all the power of my nerves and focus on just one thought, forgetting the world around me entirely. The idea is to impress on myself the object of interest so forcefully (in this case, the image of the table and book) that, when I unlock my gaze, it stands before me as if it were really there. And that's what I do. For about half a minute—as if I were "gobbling it up"—I file away the slightest details: the wrinkles in the tablecloth, its pattern, tears in the book's binding, and so on. That's enough—it's stored!
>
> Imperiously, I turn Lord to me and look deep into his eyes— or, more precisely, through his eyes, into something deep

within. In Lord's brain I implant what I have just fixated on in my own mind. Calmly, I mentally draw for him the stretch of floor leading to the table, then the table-leg, the tablecloth, and finally the book. The dog starts getting nervous, fidgets, and tries to get away. Now I give the order and a mental push: "Go!" Lord tears loose; like an automaton, he runs to the table and grabs the book in his teeth. The task is complete.[8]

Bekhterev conducted the same kind of experiments on human beings. He found easily agitated, nervous personalities especially suitable to his purposes. In tests, he sought to instill "ideas, feelings, emotions, and other psychophysical states into the psychic sphere of subjects" by "bypassing consciousness and the faculty of judgment."[9] The stimuli delivered to the brain triggered a reaction—a pattern of behavior he understood in terms of research findings on chain reflexes. Unlike Pavlov, who developed the notion of *conditioned reflex* in the same context—and did not look into the matter of "radio signals"—Bekhterev considered mental states to result from neuropsychic energy stored in ganglia (figure 3.1).

Here, parallels are evident between Bekhterev's theory of the psychic apparatus and Freud's *Project for a Scientific Psychology*. As is well known, Freud viewed the brain as an device for processing electrical signals. In his correspondence with Wilhelm Fliess, he had written:

> During an industrious night last week, when I was suffering from that degree of pain which brings about the optimal condition for my mental activities, the barriers suddenly lifted, the veils dropped, and everything became transparent—from the details of the neuroses to the determinants of consciousness. Everything seemed to fall into place, the cogs meshed, I had the impression that the thing now really was a machine that shortly would function on its own.[10]

This model, in which electromagnetic insights "precipitate," as it were, shows that Freud stood closer to Soviet brain research than is commonly held.[11] Bekhterev also viewed consciousness as a circumstance that attends certain cerebral functions: a "side-effect." He was convinced that when any given word is spoken, a thought is simultaneously transmitted; this prompted him to construct a model of the brain out of wire (figure 3.2), which he presented to the Central Committee to promote the establishment of a special laboratory for research, the aforementioned *Pantheon*.[12] Results from six different experimental arrangements gave Bekhterev reason to believe that "the mental effect of one individual on another is possible at a distance through some kind of living matter."[13] Seeking to describe this energy more precisely, he observed, "here, too, we are dealing with the phenomenon of electromagnetic energy, most likely with Hertz waves."[14]

**3.2**
Bekhterev's model of the brain

## 3.3 Mental Commands

If the living organism is capable of transforming other energy forms into a neuropsychic force, researchers supposed, then some kind of transmitter exists. On this basis, Bernard Kazhinsky was admitted to the *Zoopsychological Laboratory*, where Bekhterev  was conducting experiments with Durov. Kazhinsky, an electrical engineer, came from the field of radio research; he was known to consider human beings "radio stations" and thoughts electromagnetic waves.[15] For him, the central nervous system—including the brain—constitutes "a storage site for subtle apparatuses of biological radio communication, whose completely economical construction far exceeds the most perfect devices of technological radio communication known to us."[16]

Kazhinsky held that "living" machines for biological radio communication exist, "but they remain unknown to modern radio technology."[17] He urged his colleagues to seek out these devices through experimentation.[18] On one of his visits to the Zoopsychological Laboratory, he decided to test the influence that Durov wielded over animals on his own person. On a piece of paper, Durov wrote down the command to be transmitted, asked Kazhinsky to sit calmly, and looked deep into his eyes. The scientist reports:

> Actually, I felt nothing. Suddenly and almost mechanically, I just touched my head behind the right ear with my finger. When I brought my hand down, V. L. Durov handed me the piece of paper. To my astonishment, I read: *Scratch behind right ear*.[19]

Kazhinsky observed that he did not sense an explicit *message*—just the need to scratch behind his ear. He concluded that Durov had, in fact, "achieved a small radio transmission" into his brain, which he, in turn, "received independent of [his] conscious mind."[20]

In 1922, Kazhinsky built a screened-in construction in order to subject himself, *qua* "living radio station," to further electromagnetic investigation: a Faraday cage (figure 3.3). With this construction, Michael Faraday had demonstrated that electricity concentrates on the outside of a charged conductor; it exercises no influence on objects located within. Accordingly, Kazhinsky enclosed Durov in a box made of metal mesh.

When the door of the cage was closed, Durov, sitting inside, managed to transmit, via thought, a task to the test animal (the dog "Mars") located outside. And when the door was opened, Mars performed all the commands with the greatest precision.[21]

On the basis of this and other experiments with different versions of the Faraday cage, Kazhinsky concluded that "the nature of the phenomena accompanying the transmission of mental information at a distance is the same (electromagnetic) as in ordinary radio communication."[22]

At around the same time, the futurist Velimir Khlebnikov committed such visions of an electromagnetic world to paper in flights of technological fancy. Thus, "The Radio of the Future" (1921) declared:

> Finally we will have learned to transmit the sense of taste—and every simple, plain but healthful meal can be transformed by means of taste-dreams carried by radio rays, creating the illusion of a totally different taste sensation. People will drink water, and imagine it to be wine. A simple, ample meal will wear the guise of a luxurious feast.[23]

**3.3**
Kazhinsky's
Faraday cage

Whether in actual scientific practice or in literary experiments, the idea of brain waves connected directly to the invention of radio, which brought to life the fantasies of researchers and laymen alike.

### 3.4  Literary Circuits

(40) Belyaev's novel, *The Ruler of the World,* staged the contemporary topos of emotions transmissible by radio, which dominated science, journalism, and iconography. Like non-fictional works, it also did so to didactic and experimental ends. A large portion of the book consists of research journals offering learned reflection, for example:

> Two butterflies can communicate with one another. Yet what energy enables them to do so? ... Major chemical processes occur in the organism of living beings, and they accompany the work of muscles—and especially the operations of the nerves and brain. Hereby, distinct periodicity can be observed. That means that the nerve centers intermittently release, or radiate, ions. These ions fly off, are received by the nervous system of another being, and there we have a radio message![24]

These records stem from the hand of the German scientist Ludwig Stirner, a man unlucky in love and hungry for power. Stirner misuses his inventions in order to control others. Even before developing the plot of mass hypnosis, the novel outlines the fields of knowledge underlying telepathic communication:

> Reflexology is a science exploring the counterreactions in humans and all living beings that occur in connection with external stimuli; by this means, it explains all relations that hold between living things and their surroundings. ... A reflex is a process of transferring a stimulus to the nerves that brings a state of excitement from one point in the body to another over the center— the brain. ... A child stretches its hand to a fire. The fire burns. The effect on the skin is passed on by the nerves to the brain,

and a reaction follows from the brain: the child draws back its hand. The idea of fire and the idea of pain are now joined. Every time the child sees flame, it will pull its hand back in fear. What scientists call a "conditioned reflex" has emerged.[25]

This transposition of scientific doctrine into literary discourse underscores what interested contemporary researchers. Bekhterev and Pavlov did not limit their inquiries to the physical reactions provoked by pain. Rather, they made it clear that all new cultural-technical processes likewise involve reflexes:

> What is revolution in general? It is liberation from all … inhibitors. … It is complete absence of restraint. There were laws, customs, and so forth. All of this has now come to naught. The old is gone, the new still does not exist. Inhibition is eliminated, there remains only excitation. And this produces all possible excesses in the realms of desire, thought, and behavior.[26]

The insight, then, is that knowledge emerges through a process, in stages; learning should be understood in terms of reflexes. In this context, novels represent elements of a system of reading that not only possesses the power to disseminate science, but itself constitutes an experimental *praxis*.

In Belyaev's novel, Stirner observes that he has dissected more than a thousand brains but never found the mind (*um*).[27] When he recognizes the power he can exercise over dogs, he decides to apply the same methods to human beings. In particular, he directs his mental power at Elsa Glück, the woman who has rejected his amorous advances. The "secret" to wielding control is "picturing, with utmost precision, each of the actions the person to be influenced should perform."[28] The victim becomes the slave of Stirner's emotions; her readiness for self-sacrifice mirrors his narcissism. "More and more, she's turning from a living human being into an automaton."[29] In a parallel development, the populace at large becomes concerned about increasingly frequent cases of robotic behavior exhibited by the city's residents. Ultimately, the hypertrophic proliferation of discourses makes a matter of suspicion a certainty: Stirner is using science to pursue comprehensive mental control.

Time and again, the book gives space to suspicions voiced by Stirner's rival about telepathy—the mass manipulation the monstrous hypnotist is pursuing:

> I'm no scientist, and I don't know how Stirner managed to get his own thoughts into other people. But I think his power is limited to his circle of acquaintances. I can say as much because only here did I feel how, with time, I started to free myself from hypnosis, to "demagnetize" the charge I had received before my departure.[30]

Passages such as this one function as rhetorical fields of radio waves, where all manner of speculation about telepathy occurs and the possibility of occult and suprasensory channels is broadcast.

Ultimately, however, criminological discourse yields to a gesture of scientific enlightenment. When efforts to assassinate Stirner prove bootless, the inhabitants of the "irradiated" city find a weapon that works: a scientist and researcher who is "just as strong" as Stirner. This individual describes his insights to another character:

> To my astonishment, I found a series of closely related analogies between the design of a nervous system—like the brain—and that of a radio station. Parts of the brain play the role of a microphone, detector, and telephone. The fibrillar strands of neurons have a twist at the end that calls a wire-like spiral to mind; here, induction occurs. From the perspective of physiology, it is interesting that even the professor of physiology with whom I was working proved unable to provide a satisfactory answer about the function of this coil. But in light of electrical engineering, there's a logical meaning. Nature obviously created this coil to amplify electrical currents. Our bodies are even equipped with Round's light-emitting diodes—they're the

valves of the heart. The source of cardiac energy corresponds to an accumulator, and the peripheral nervous system to grounding. In investigating the human body from the perspective of electrical engineering, I came to the conclusion that our body represents a complex electrical apparatus, a whole radio station, which is able to send and receive electromagnetic vibrations. If you please, here's the drawing.[31]

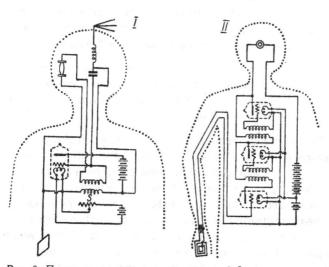

Рис. 9. Первоначальные схемы передающей I и принимающей II биорадиостанций нервной системы человека.

**3.4**

**Radio-person, following Kazhinsky and Kachinsky**

A diagram (figure 3.4) accompanies this lengthy passage abounding in technical jargon. Together, the text and the image provide detailed information about the relationship between human beings and radio—a shared mode of functioning. Significantly, such data does not come from the laboratory reports of Bernard Kazhinsky, the real-life telepathy researcher. Instead, it comes from a character: the Soviet engineer *Kachinsky,* whose name differs from that of his counterpart by only a letter. Belyaev took up the theories and practices of contemporary research on telepathy and put them to work in his novel.

Science is directly transposed into literature: not only do Kachinsky and the trainer Dugov stand for the historical figures Kazhinsky and Durov; the professor of physiology in the novel represents Bekhterev. These characters relay scientific insights, now arranged in didactic form for narrative purposes. For example, Kachinsky describes an experiment central to the plot:

> We put a dog in front of the cage, and Dugov inside. When the cage wasn't grounded, the dog successfully performed Dugov's mental commands. But as soon as it was, the channels of influence no longer reached the dog.[32]

The contrivance winds up being an important part of the weapon Kachinsky uses to fight against his telepathic antagonist. Needless to say, this aspect of the work also represents the point where experimentation switches over from actual scientific research to a process of *construction* that is literary in the proper sense. The differences between real-world tests and their fictional extensions are not marked (or significant) as such; instead they serve the purpose of heightening suspense and engagement on part of the reader—plugging him or her into a broader cultural network all the while.

Kachinsky, having studied the electromagnetic waves that the brain emits, can determine their length and frequency; by this means, he is able to reproduce thoughts mechanically. That said, he needs a *brain machine* to complete his weapon. It has not been built yet:

> We have limited ourselves to tests of sending thoughts to animals at short distances. But my "brain-machine" can be made with modern technology. ... If [thoughts] are amplified with transformers, then the thought waves will flow like ordinary radio waves and be received by human beings.[33]

Kachinsky's device consists of "an antenna, an amplifier with transformers, cathode lamps, and an inductive charger."[34] Needless to say, the components of the brain machine connect directly with achievements in radio technology charted from the beginning of the twentieth century on.

The following words were not written by Belyaev, but by the radio pioneer Alexander Popov (1859–1906):

> Effect at a distance can be accomplished by two very different methods: through a strong charge that, surfacing periodically and then waning, stimulates the electrostatic field; or through the electromagnetic effect of intermittent or alternating current.[35]

Thomas Alva Edison had developed these operations; in turn, Nikola Tesla perfected them. Credit for the further evolution of wireless communication is a matter of controversy involving Popov and Guglielmo Marconi.

In 1896, Popov had constructed a *Device for Detecting and Recording Electric Vibrations* (figure 3.5)[36] that could capture electromagnetic waves from the atmosphere. On 24 March of that year, he demonstrated the wireless transmission of signals at a distance of 250 meters to members of the St. Petersburg Academy of Sciences. In June, Marconi patented a device in England that had the same design as the one already outlined in his rival's publications. Popov responded with a series of statements in the national and international press defending his priority and rights. He received due honors for his innovation at the Paris Exhibition

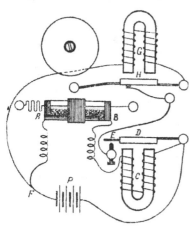

Рис. 1. Схема первого в мире радио-
приемника, изобретенного А. С. Попо-
вым и названного им «грозоотмет-
чик».

**3.5**

**Popov's design for
a radio receiver**

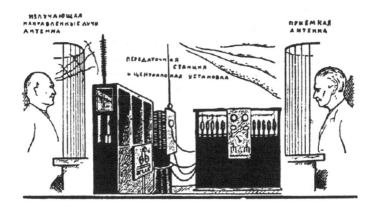

in 1900, but Marconi still counts, at least in the public eye, as the inventor.[37] Kazhinsky credited the "Russian inventor of the radio" with having inspired him to look for electromagnetic vibrations in human beings. Even if Popov himself doubted that the body possessed such properties, Kazhinsky took this same skepticism as proof that he had "chosen the right path."[38]

In 1924, Hans Berger invented the electroencephalograph, which made it possible to record the *physical energy* of the brain.[39] The development could only have made Kazhinsky redouble his efforts: the device's sensitive measurement amplifier, able to register signals in an order of magnitude from 5 to 150 μV, permitted brain activity to be determined by recording current fluctuations at the top of the head. In Belyaev's novel, Kazhinsky's counterpart Kachinsky uses these same components to construct his new weapon: the *brain machine* and Faraday cage (reworked as a "gun shield") are what finally enable him to take up the fight against Stirner (figure 3.6).

The drawing in the book resembles Gulyaev's model of communication (figure 1.1), but it differs, too. Most significantly, the polarity of the figures in the novel is not the same—at least until the end. In the course of mental combat between politically unequal parties, the Russian researcher Kachinsky proves vastly inferior to his German antagonist. Stirner "paralyzes" the "brain centers governing equilibrium" in his opponent (the same centers, Kachinsky notes, which Bekhterev has studied assiduously[40]). The Russian manages to heal himself homeopathically, only to find himself forced by the German's mental power to return to his quarters, where he then lies, incapacitated.

**3.6 (left)**
Thought-war

## 3.5 The Death of Consciousness

The novel's material dimensions and the recursive discourse-formations it establishes *prescribe* technological "paranoia" to the reader, which only gains in persuasive force as the narrative proceeds. Its technoid terminology and infographics make as much plain. More still, inasmuch as the book—the symbiosis of technological fact and cultural instruction—*broadcasts* detailed information about influencing apparatuses and their relation to familiar spheres of life, it affords the public the practical "demonstration" of a controversial science. The system the fiction elaborates does not require that thoughts be verbalized for political restructuring to proceed; linguistic signs lose their meaning more and more.

In historical context, the correlative project of making a New Man endowed with telepathic destiny [*Sendungspotenzial*] extended to everyday life and society via real-world experiments seeking the same—no matter whether they actually worked or not. The model of communication Belyaev took up in, and disseminated through, *The Ruler of the World* attaches singular importance to e*mission* and trans*mission*. The very possibility of such an exchange between fact and fiction points, over and above the book's technological themes, to the process of transfer, interference, and storage identified by Claude E. Shannon.[41] The "significant" feature of communication, Shannon observed, "is that the actual message is one *selected from a set* of possible messages."[42] Based on probabilistic Markov chains, his theory posits that language admits calculation—in other words, that nothing distinguishes linguistic operations from technological modes of communication [*Übertragung*].[43]

The novel concerns the effects of a source of interference coursing through the ether. In Soviet discourse networks, fictionalized experiments, which amount to a reactive medium for propagating scientific findings, both transform standing insights and call forth new social and political practices. When telepathy is represented, the signifying practices of theoretical research and tests fuse with literary strategies. Here, believability is not conceived as an object that is simply given; rather, fiction provides a statistical, culturally scaled, and *descriptive* account of phenomena. In this context, decoding science translates into the coding of consciousness. The process of laboratory experimentation holds an exceptional epistemological status in that it makes its own radio-technical arrangement into the scheme for electromagnetic thinking on the part of the reading collective.

That said, the novel's psychography implicitly and explicitly elaborates a media theory that hypostasizes a closed order of reception-, production-, and transmission technologies. As such, it is only boundless love for Elsa that can induce Stirner, the commander of thoughts, to abandon his project when he is no longer able to bear manipulating his beloved. The outcome of the plot hinges on the telepathic engineering of a subject that is mature, autonomous, and free to think. Stirner is the "guinea pig," and he performs the test on himself:

> Stirner must die. I [Stirner] have given a command to my thought-transfer machine. I set it to the highest capacity. At precisely one o'clock in the morning … , it will broadcast this command: from Stirner to Stirner. He will lose consciousness. He will forget everything that has occurred in his life. He will be a New Man, charged with new consciousness. This will be Stern. Stern will go where Stirner ordered him to go. And Stern will not even suspect that Stirner, locked up in the iron cage of his unconscious, is whiling away a miserable life. It is death … the death of consciousness.[44]

Symbolic practices, once set in motion, operate independently and bring about hyperreality—a second world of active simulation—which, as the sum of *ambient dispositives,* feeds into (mental) representations, needs, desires, and perception.

From this point on, no *outside* exists for Stirner; the external world is also stuck in the prison of simulation—what Baudrillard would later call "hallucination of reality,"[45] where the greater mass of humankind idles away. Now, only Stern will live: a man who "naturally" occupies a single site—the only place he can live out his newly *set* way of thinking. This is a land where all men are equal; everyone is a star [*Stern*].

Belyaev's *Ruler of the World,* besides depicting the transformation of science, participated in the process of secularization affecting society as a whole; in this capacity, it drew new borders for metaphysical (or religious) dispositives of interpretation. For all that, the questions it posed to natural laws, which had counted as simply given until then, did not necessarily turn against beliefs [*Glaubensvorstellungen*]; on the contrary, such questioning provided parascientific, auxiliary argumentation buttressing transcendental schemes of explanation.[46] Neither in fiction nor in reality did the public sphere lie beyond the grasp of *electromagnetic* thought-rays—in a space screened off from scientific endeavor. Instead— and as occurs in Gulyaev's drawing mentioned at the outset (cf. figure 3.7, detail of figure 1.1)—the public sphere continued to provide the site where interference between scientific and extra-scientific discourses switched over into productive exchange.

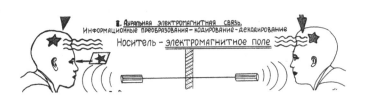

**3.7**
Auratic-electromagnetic switching

Accordingly, in the discourse network of Soviet telepathy, experimental praxis did not arrive at solid data so much as it brought forth literary forms of science propelling this same science onward. Inasmuch as the graphic traces of electromagnetic faith changed into writing which the broader public read, literature did not tangle telepathy into a knotted coil of measurements and curves. Instead, it spun them out. It pursued didactic aims and effected enlightenment: describing the future in order to inspire belief. In turn, this inscription of telepathic research into fiction echoed utopian, political visions of power.

(51)

At the end of *The Ruler of the World*, Moscow is described as a laboratory of peace and quiet. Communication takes place here and now. This calm state contrasts with the "noise" that necessarily attends the mail, telegraphs, or telephoning:

> Moscow has become a city of sublime silence. We almost no longer speak with each other, since we have learned to exchange thoughts. How unwieldy and slow the old way of talking now seems! Perhaps we'll forget how to speak entirely over time. Soon, mail, telegraphy, and even radio will belong to the past [*my sdadim v arkhiv*].[47]

Clearly, no one has any secrets in this new social system, for everyone is *synchronized* [*gleichgeschaltet*]. What the *one* thinks is immediately present in the head of the *other.* Communication is machinic and operational—perfectly transparent in a perfectly transparent society.[48]

The passage also makes it clear that so long as telepathy cannot be demonstrated to exist, it must be devised. Precisely this is what literature does through the "magic" of the word. That is, the text does not just mix together science and fiction; rather, it reenacts experimental practices and reactivates their premises, which are based on actual laboratory work. Where the novel presents fictional elements, it employs the same procedures that contemporary Soviet brain research was using (or anticipating).

The book's telepathic discourses attest to the contemporary boom in *mental* labor. The epistemic phenomenon was not restricted to laboratories; it also haunted the office of the general secretary. Telepathy was not just a matter of experimentation detached from life—it extended to everyday existence and transformed the public sphere into a field of research for synchronizing the thoughts of the masses. This social laboratory had the goal of reconfiguring the minds, indeed the collective psyche, of Soviet citizens, who "do not even suspect anything about the secret files of the unconscious, into which the mighty concentration of thought has locked away all that was fearsome and dangerous to humankind."[49]

Bekhterev might have known that, in the "City of Silence," the very idea of diagnosing "paranoia" in the ruler would produce fatal results.

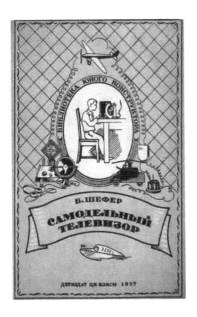

**4.1**

Homemade television

# 4

# ◾ Televisual Passions

The psychic and social disturbance created by
the TV image and not the TV programming,
occasions daily comment in the press.
**Marshall McLuhan**

## 4.1   Greetings from Afar

"Dear Comrades!" began the lead article of *Pravda* on 29 November
1981. The headline addressed workers of the Soviet television,
communication, and broadcast industry. The piece continued: "I
congratulate you wholeheartedly on the occasion of a great event —
fifty years of regular televised broadcasts!"[1] The author was none
other than Leonid Ilyich Brezhnev, fourfold hero of the Soviet Union;
at the time of writing, he had held the office of general secretary
of the Communist Party of the Soviet Union for seventeen years.
Now, in the public eye, he was appealing directly to "journalists,
editors, directors, camera operators, engineers, technicians, and
workers" — "today, because of mass communication and propaganda," they faced "important tasks." The Soviet Union com-
manded the largest television system in the world; by "constantly
developing and modernizing," it was "solidifying and spreading
the material, technological foundation"[2] for Communism.

The general secretary might have offered his felicitations half a
year earlier. In fact, the first television broadcast on Soviet territory
occurred on 29 April 1931, even though regular programming did

not start until 1 October.[3] Either way, Brezhnev's congratulatory words harbored a paradox: although news and propaganda were routinely broadcast starting in 1931, the Soviet industry of the time had not yet produced a single television set.[4] The only devices in existence had been built by amateurs and hobbyists following plans patented by Hovannes Abramovich Adamian.

56

Specialized journals, daily newspapers, and small brochures (figure 4.1) had helped Homo sovieticus set up television laboratories in the home.[5] The publication of television designs served not only to convey complex electrotechnical information to citizens; equally, it "plugged them in" and activated newly gained insights.[6] The goal was to make the people perk up its antennae in order to receive the transmissions of New Soviet Thought. In historical perspective, the citizenry's encounter with technology—anchored in the visibility of the apparatus[7] and based on designs and patents for "seeing machines"—unfolded as a process of auratic-electromagnetic switching.

## 4.2 "Intellectual Property"

In 1879, Eadweard Muybridge invented the zoopraxiscope, a device for projecting a continuous sequence of photographic images; until now, they had been fixed and immobilized on glass slides. George Carey also published his models for effecting electrical transmissions over distance. That same year, on 17 February, the pioneer of Soviet television, Hovannes Abramovich Adamian (figure 4.2), was born. Adamian enjoyed a comfortable childhood as the scion of a merchant family in the Armenian city of Baku, on the west coast of the Caspian Sea. As a young man, he went to Zurich and Paris to study physics, which triggered his interest in developing ways to transmit and receive images. Adamian then spent a good thirty years conducting research in St. Petersburg, where he died of lung cancer on 12 September 1932.

There are many reasons Adamian settled in St. Petersburg. The
city had been the capital of modern science in Russia from the
eighteenth century on.[8] Here at the Academy, one might encounter
Dmitri Ivanovich Mendeleev, the inventor of the periodic table of
elements, Pavlov, or Bekhterev (whose dog-training sessions Adamian regularly attended). Most significantly—and especially after
Popov's successful radio experiments—St. Petersburg became
home to wireless telegraphy, which not only promised that images
would soon be transmitted, but also suggested that the reality of
telepathy and clairvoyance would be proven scientifically.[9] It was
fitting, then, that television should be born here.

Adamian left the city on the Neva rarely: to spend a vacation in
his native Baku or for travel to Berlin in order to discuss new devel-
opments in phototelegraphy with colleagues and competitors—
and, if need be, to register a patent first.[10] Popov's case had cer-
tainly made it clear to him that czarist Russia showed little concern
for protecting intellectual property: the inventor of wireless telegraphy had been forced to surrender credit for his innovation to
Marconi. (This was yet another reason that Lenin's decree, mentioned above, set an end to the standing system of privileges and
handed profits over to the state.) Needless to say, Adamian also
patented his inventions, which were all based on electrical power,

in the Soviet Union.[11] Perhaps he believed in the new regime, which sought to realize communism by way of the equation "Soviet power plus the electrification of the whole land"—indeed, devices he invented for a better harvest and healthy bodies (figure 4.3) were part of this program.

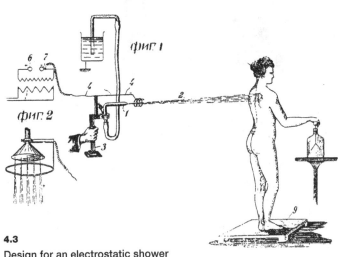

**4.3**
Design for an electrostatic shower

## 4.3 Images, Holes, Signals

From the outset, Adamian knew that phototelegraphy, which had existed since the nineteenth century, needed to be combined with the cinematographic principle of image sequencing in order to make television possible. On his visit to Berlin in spring 1907—when he surely discussed these technologies with telecommunications pioneer Arthur Korn[12]—he submitted his Device for the Recording and Reproduction of Electrical Images and Image-Sequences to the Imperial Patent Office. An account of its utility precedes the technical details:

The invention concerns a device for the retention and repeated reproduction of images or series of images, which have been carried at a distance by means of known instruments for electrical conduction. The device requires a transducer that dissolves the image or series of images into an arbitrary quantity of dot-shaped elements; it produces a sequence of current fluctuations in the circuitry, whose amplitudes are approximately proportionate to the intensities of light in the pictorial elements conveyed in succession.[13]

In essence, Adamian's invention consists of four elements: a transmitter that can transform images into electricity, a receiver with fully electrical switching networks, recording equipment, and a component for transforming the current back into images. The schematic design of the transmitter is reproduced in the first illustration accompanying the patent document (figure 4.4): the image to be transmitted (1) is captured by a camera lens (2) and projected onto the screen (3). Directly behind (or in front of) the screen "there is a disk (4) that can be turned, whose edge is perforated with holes (5)."[14] (Adamian indicates that it does not matter whether the disk is placed in front of, or behind, the screen.[15]) The disk's functionality is described as follows:

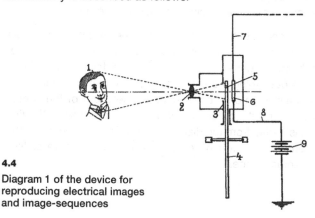

**4.4**

Diagram 1 of the device for reproducing electrical images and image-sequences

The holes are arranged in familiar fashion along a spiral, so that the hole visible in the drawing stands closest to the edge of the disk, and each following one a little closer to the middle axis. Behind the disk (4) a system of selenium cells (6) has been arranged in a chessboard pattern; its size matches the size of the image falling onto the screen (3), and its cells correspond to the dot-shaped pictorial elements that are to be transmitted. All selenium cells are switched in parallel between circuits (7) and (8). Circuit (8) is switched with an energy source, whose free pole is connected to the ground.[16]

The phrase "arranged in familiar fashion" points to the "technological nucleus of television"[17]—albeit without naming it. Transforming images into currents was the starting point for developing television technology.[18] The disk to which Adamian refers was an essential element of the electrical telescope Paul Nipkow had patented in 1884; it represents an essential component of almost every television project to follow.[19] On one end, the Nipkow disk breaks down the image in terms of width and height; this information is transmitted in a series of individual points similar to Morse code. On the receiving end, the images are reassembled by means of another, synchronized Nipkow disk. As the disk spins, one dot follows another in a (relatively) straight scanning line:

When a point at the left edge of the screen window disappears, the next point in the spiral, shifted by a row toward the center, reaches the right edge of the screen window; it immediately begins describing the following line. The angular distance of the holes therefore corresponds to the width of the screen window, and the difference between the distance of the first and last holes from the disk's center defines the height of the screen window, which necessarily assumes a more or less trapezoidal shape.[20]

Because the disk rotates rapidly (about fifteen times per second), the sluggishness of the human eye guarantees that the observer cannot perceive individual dots; the picture draws itself, so to speak. This same process underlies the transmitter outlined in Adamian's first figure: as the perforated edge of the disk cycles along, "individual punctiform elements of the image are allowed to pass, which fall on selenium cells"[21] arranged behind the screen; in turn, the resistance of cells changes in response to varying degrees of exposure, which entails a corresponding fluctuation of the current.[22]

Although Jacob Berzelius discovered selenium in 1817, it was not until 1873 that British engineer Willoughby Smith and his assistant Joseph May demonstrated the element's photosensitive properties.[23] Nipkow recognized that selenium's photosensitivity could be used to effect voltage pulses. Adamian's design also registers these fluctuations in the receiver (figure 4.5); a "Siemens oscillograph" (10), the patent document specifies, switches between the section lines. What Friedrich Kittler has written apropos of Nipkow's invention holds here, too: "the discrete line form of the book or telegram is imposed onto images."[24] Light from the mirror (11) of the oscillograph (10) is projected onto photographic film (26) being drawn past continuously. In this way, the film registers a curve "containing all elements of the image, split into temporal or spatial sequences."[25]

Following the description of mechanisms for sending and receiving (which are not new), the patent document comes to Adamian's own invention (figure 4.6). The device is meant to reproduce the original image following the photographic curve as quickly and distortion-free as possible.

It consists of an apparatus (27) for producing rapid electrical discharges, specifically a Ruhmkorff spark generator, a Leyden jar, a spark gap, and an induction coil, moreover a rotating punch disk (28) with a viewfinder (29) and a Geissler tube (30).

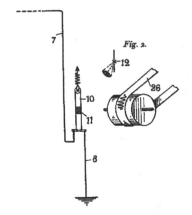

*Fig. 2.*

**4.5**
Diagram 2

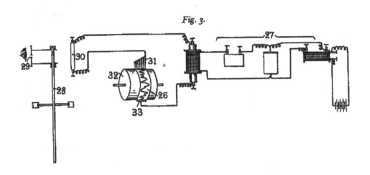

*Fig. 3.*

**4.6**
Diagram 3

A device (31–33) is switched into rows between the spark generator and the Geissler tube. It consists of a series of resistors arranged by size (31), each of which is connected to the circuit at one end and terminates in a point at the other. The points are arranged like a comb over a cylinder (32); on this cylinder, the paper strip or film (26)—after the curve photographed on it has been made conductive—is wound. A given frame of film has a corresponding electrode in the circuit.[26]

(63)

In order to transmit the image, "the punch disk (28) is made to spin and the spark device (27) is switched on."[27] The electrical current flows into the Geissler tube and then through the row of resistors; it "can only leave a tooth of the comb (31) matching a curve-sequence, since only the curve itself is conductive."[28] Image-generation occurs as the cylinder turns, whereby changes in the coordinates of points on the curve switch into resistance changes in the circuit. "Changing intensity in brightness of the current in the Geissler tube"[29] corresponds to varying levels of resistance:

If the punch disk (28) simultaneously rotates at a speed in proper relation to the rotational speed of the cylinder (32) and is in correct phase with the curve, the eye sees the image, or image-sequence, that has served to construct the curve.[30]

These recording practices and strategies, which were developed and verified through models for television, stand in immediate proximity to another kind of experimentation that was fashionable at the time.[31] In both pre- and post-revolutionary Russia, spiritist séances were the order of the day. Not just credulous parties, but also leading figures in the fields traversed by occultism (e.g., psychology, physiology, physics, and radio technology) were represented. According to records, "Comrade Hovannes Abramovich Adamian" participated in sessions on several occasions.[32]

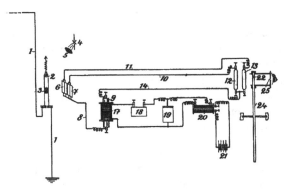

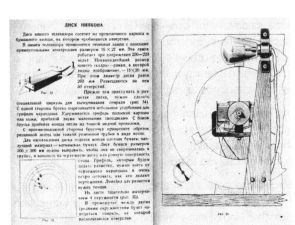

## 4.7

Instructions for converting
fluctuations of a light beam
emitted by the mirror of an
oscillograph

## 4.8

Instructions for constructing
a Nipkow disk

## 4.4 Lights, Angles, Intensities

Another invention Adamian submitted to the Imperial Patent Office during his time in Berlin was for an electric television. The documents outline the design for an apparatus (figure 4.7) connected to devices "in which the varying brightness of individual points within an image is used to produce fluctuations of current in an electrical circuit."[33] His Device for Converting Local Fluctuations from a Beam Emitted by the Mirror of an Oscillograph into Luminance Fluctuations of a Geissler Tube serves "to reproduce," via electrical current, "the image seen at the point of origin at the point of reception."[34]

Here, too, the matter involves arranging a number of selenium cells in a chessboard pattern and exposing them to flashes of light to break the image down into a series of temporally sequenced current fluctuations registered by an oscilloscope. In this process, the mirror of the oscilloscope initially turns the current fluctuations into "spatial fluctuations of a light beam (of arbitrary intensity) emitted from a suitable source."[35] Adamian's innovation lies in the next step:

> The invention is a device through which the angle fluctuations of this beam of light are transformed, in the receiver-apparatus, into intensity fluctuations of one or more light sources—for example, Geissler tubes—which then can be arranged, in familiar fashion, so that the image taken up by the transducer is passed on to the receiver, for the eye or photographic camera.[36]

The design calls for using two light sources of different color, preferably white and red; they are to be arranged in such a way "that the lighter tones are reproduced as white by one light source, and those that are less bright as red or reddish."[37] Adamian makes this proposal in view of the human face, "whose natural colors are to be imitated, to a certain degree, by these color gradations."[38]

This apparatus, intended to transmit images more precisely than the earlier one, operates as follows:

When current fluctuations flow through the circuit (1), the mirror (3) and the light beam emitted by the lamp (4) are diverted accordingly. Thus, if the mirror is set in such a way that the beam, in off-position, does not shine on any of the selenium cells, the apparatus, as depicted, allows for five different degrees of brightness to be transferred. In keeping with whether the beam shines on the first or the second cell of the first or second cell-group, the resistance of the entire system of cells changes; for instance, it becomes smaller, the more the beam moves rightward. Conversely, the current flowing through and lighting up both the Geissler tubes increases or decreases in proportion to the resistance of the cell-system.[39]

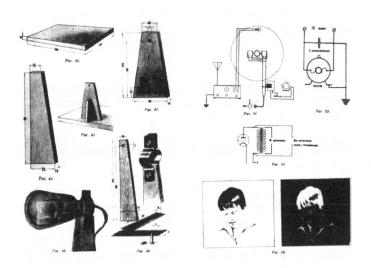

**4.9**
Televisual designs

The color of the "light emitted" changes because the resistance of unexposed cells blocks the current; the white tube remains dark until the beam of light reaches the group of cells on the right.

Once documents were filed, the public had access to designs. The Soviet patent system allowed the state to make individual intellectual property redound to general benefit—to go straight to the people, without detour. This is why many construction guides for television appeared in the print medium: texts tinkerers could use to construct their own Nipkow disk (figure 4.8); designs for setting up tubes, lights, and selenium cells (figure 4.9), and so on. Such circuit diagrams enabled them to make their own device for receiving the uncanny signals, forever switching between black and white, that came to fill the ether during the Technological Age.

(67)

## 4.5 On the Air

In the night from 29 to 30 April 1931, Adamian's lab technicians conducted the first wireless television transmission in the Soviet Union. They broadcast both the equipment and objects necessary to construct and work seeing-machines and the contraptions' operators (figure 4.10)—i.e., themselves—as a host of spectral phenomena.[40] The quality of the images was minimal inasmuch as the televisual scans comprised just thirty rows and were about the size of a matchbox.[41] All the same, the forms and contours of what had been sent out into space could be clearly distinguished.

The next morning, *Pravda*—"The Truth"—reported the experiment's success and drew readers' attention to a public test scheduled to occur the following evening:

> Tomorrow, for the first time in the USSR, an experimental televisual message (seeing-at-a-distance [*dal'novideniya*]) will be transmitted over the radio. Over the short-wave station RV-1 at the All-Union Electrotechnical Institute (Moscow), a living face and a photograph will be transmitted at the wavelength of 56.6 meters.[42]

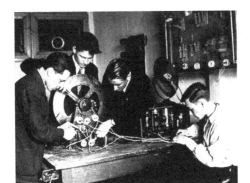

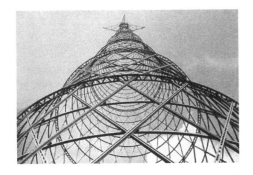

**4.10**

Laboratory technicians
preparing the device for a
broadcast session

**4.11**

The Shukhov Tower from
Alexander Rodchenko's
perspective

Before television could broadcast propaganda and fetter minds, the device had to be calibrated to the masses. As opposed to the tuning process in capitalist Germany, where uninhibited "bathing nymphs" lured the male gaze beyond the camera, as it were,[43] test images meant to monumentalize communism were transmitted. Thus, the article alerted readers to the fact that, after the previous night's emission, the broadcast the following evening—the night before May 1—would feature a second image in addition to the picture of a human being.

(69)

The image in question was a photograph: a hyperbolic antenna soaring to the height of 150 meters, which had been broadcasting radio signals for the last nine years. Even though plans in 1919—before the world financial crisis and the Russian Civil War—had foreseen a height of 350 meters, Vladimir Grigoryevich Shukov's design stood as an icon of Soviet progress in the public eye. Alexander Rodchenko captured the constructivist majesty of the Shukhov Tower (figure 4.11) in a picture that radiated its own aura, which electromagnetic waves now amplified and diffused.

**5.1**

Listen, Moscow is speaking!

# 5

# Cybernetic Circuits

Pure, simple declaration without justification or
proof is a sure means of instilling an idea in
the human soul.
**Gustave Le Bon**

## 5.1 Soviet Ether

When Soviet rule began, both scientists and the new intendants of
power were convinced that mental energy was the same thing as
electromagnetism. It was only a matter of time before communist
society would no longer depend on wires in order to hear Moscow speak (figure 5.1).[1] In his visions of a radio-controlled world,
Khlebnikov did not present the advantages of the new medium
only in the abstract; he celebrated their social utility, too: "Doctors
today can treat patients long distance"—he wrote in his plan for a
society suffused with electromagnetic waves—"through hypnotic
suggestion. Radio in the future will be able to act also as a doctor,
healing patients without medicine."[2]

But even though leading engineers, radio technicians, and phys-
iologists conducted experiments that demonstrated electromagnetic emissions of thought,[3] statistical checks put such affirmations
into question.[4] Subsequently, during the Stalin era, the fictions
promulgated in the media no longer followed enlightenment-vision-
ary ideals or the technological designs of the avant-garde.[5] Communication networks continued to show a marked tendency toward

dematerialization and hypertrophic mythologization, but now Stalin and the dialectic of historical evolution he incarnated took center stage: the head of state embodied all possible influence apparatuses, whether material or immaterial (figure 5.2).[6]

ПОД ВОДИТЕЛЬСТВОМ ВЕЛИКОГО СТАЛИНА—ВПЕРЕД К КОММУНИЗМУ!

**5.2**
Under the guidance of the great Stalin—
onward to Communism!

And so, a curious, archaizing dynamic arose. The tools of modernity brought forth premodern culture, technology produced pretechnological forms, and the media generated unmediated experience. Accordingly, Stalin-era dispositives may be viewed from two angles: both as allegorical signs of all-encompassing Soviet power and—in an altogether concrete sense—in terms of control via a media system overseen by the general secretary acting in an official, bureaucratic capacity.[7] This configuration would prove decisive for the representation of the media in Soviet propaganda.

**5.3**
Zhukov and the plan
of attack on Berlin

The famous "wall scene" in Mikhail Chiaureli's *The Fall of Berlin* (1950) is exemplary in this regard. A discussion is taking place between generals—including Zhukov—about the strategy for conquering the German capital (figure 5.3). They study a map marked with red arrows representing various army divisions; all the arrows terminate in a black point: Berlin. A traveling shot heightens the symbolism of military artistry coordinated by means of immaterial communications. Then, without a cut, the shot passes through the wall (figure 5.4) and shows another gathering—not on the front, but at Yalta (figure 5.5). Now, Allied heads of state and officers listen attentively to the Soviet general secretary, who provides precise information about the operation. When Stalin speaks, everyone listens. He requires no apparatuses at all in order to communicate.

Stalin, then, incarnates the medium of all media; his omnipresence is guaranteed even without material supports.[8] The intersection of symbolic hegemony and pragmatic bureaucracy represents the hypostasis of a new, thoroughly networked society with nothing to hide—and not just from its ruler. This interpretation is lent further cogency inasmuch as the general secretary continued to embody the spirit of telepathic communication after death. When

**5.4**
The wall between the front
and Yalta

**5.5**
Stalin and the Allies

Stalin had passed from this world, research aimed to develop the means to heat up Cold War channels in Western minds. The right device, propaganda held, could spy out industrial secrets or locate hidden facilities and bases.[9] The concrete demonstration of such apparatuses—or how they might be constructed—played a relatively slight role. The real point was that such claims would penetrate the unconscious mind of the masses, domestically and abroad.

(75)

## 5.2  The Clinic of the Future

"Imagine the polyclinic of the near future," Gulyaev wrote in a 1969 article, "Biological Communication at Work" (*Biologicheskaya svyaz' deystvuet*):

> A patient enters the treatment room. The apparatuses register the electro-auragram of his brain, heart, nerves, muscles, and internal organs; they send the information gained to an electronic diagnosis machine that, after it has determined the illness, indicates the proper treatment. All this occurs in a few seconds; the patient needn't even undress.[10]

This contribution appeared in a journal of popular science, Technology for the Youth (*Tekhnika—molodezhi*), devoted to the theme: In the World of Living "Radio Stations" (figure 5.6). Other pieces by researchers at the Laboratory for Physiological Cybernetics—which Gulyaev directed—reported on findings from the last twenty years. The premise for investigations was that human organs—the heart, muscles, nerves, and, last but not least, the brain—are electrogenetic; as they perform their functions, they bring forth currents that spread throughout the entire body, yielding an "inner, electrotonic field."

The articles discuss the "electrical landscape of nature": how plants and trees cast electrical shadows, which screen off electrical fields; meanwhile, wings, hair, and fur—"experimentally proven" to be electrostatic microphones—serve as carriers for electrical

**5.6**

In the world of living
"radio stations"

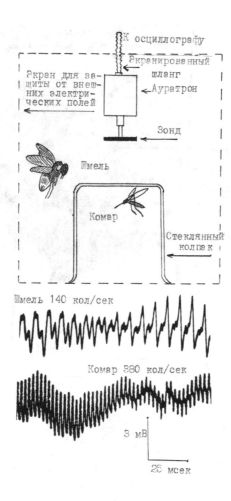

К осциллографу

Экранированный шланг

Экран для защиты от внешних электрических полей

Ауратрон

Зонд

Шмель

Комар

Стеклянный колпак

Шмель 140 кол/сек

Комар 380 кол/сек

3 мВ

25 мсек

**5.7**

**Aurathron-measurements
of a bumblebee and a
mosquito**

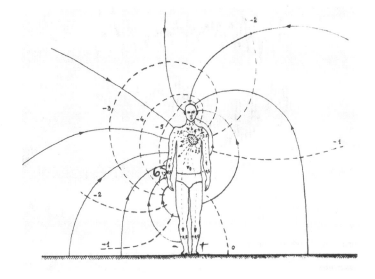

voices. Accordingly, researchers have succeeded in recording the auragrams of an array of creatures: a bumblebee, a honeybee (figure 5.7), a wasp, a fly, a mosquito, and, last but not least, a human being. Discovery of the electrical field of the human nervous system, which can be measured at a distance of 25 cm, holds phenomenal implications.

This field exists only for a thousandth of a second (for as long as the impulse runs down the nerves). If one considers that there are more than four million neuronal channels in our body and the activity of impulses is always changing, one can imagine how complex the whole nerve field of a Homo sapiens looks.[11]

Not only have the technicians working under Gulyaev (figure 5.8) concluded that living organisms actively generate electromagnetic waves; more still, the aura field—which, like all electrical fields, can be charted in terms of current density—possesses vector properties. On this basis, it is possible to determine the activity and mechanism of the heart and brain (figure 5.9); the experimental results connect directly to contemporary efforts to organize physical and mental currents cybernetically.

Earlier in the century, Stalin's doomed diagnostician Bekhterev—whose thoughts were then "transmitted" to his student Gulyaev—had explored traditional ideas about transmaterial communication and intervention.[12] He set his research apart from shamanistic methods of healing indebted to hypnosis by pointing to the clinical reality that suggestion produces: *contagium psychicum*.

**5.8**
Pavel Gulyaev in front
of a map for electrifying
the Soviet Union

**5.9**
Aura field with vector
properties

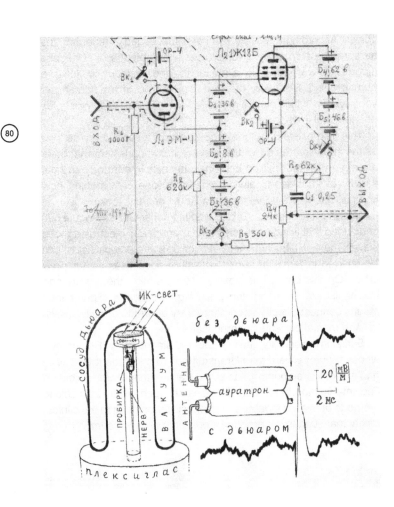

**5.10**

Circuit diagram and measurement
of an Aurathron

Just as physical germs of infection produce massive effects and can prove ruinous, far beyond the individual scale, for entire population groups, so, too, do psychic agents of contagion tend to spread; they are active everywhere and conveyed by words or gestures, through books and newspapers. Psychic "microbes" are all-pervasive and capable of developing under all conditions; wherever we may be, the danger of psychic infection exists.[13]

Bekhterev viewed expressions of the soul [*seelische Äußerungen*] as the effects of neuropsychic energy stored in the ganglia. Since the living organism is capable of transforming other kinds of energy (warmth, light, and so on) into neuropsychic energy, Bekhterev sought to understand how thoughts circulated as a kind of radiation.[14] This same energy is precisely what Gulyaev now, decades later, thought he could finally measure.

By his own account, Gulyaev—after studying physics and biology and before devoting himself to auratronics—had fought against the German Wehrmacht. His background in physiology benefited operations on the rear front, at Yelabuga on the Kama River. Here, he combated every soldier's "inner enemy": fatigue. The research he and colleagues conducted focused on the power of fenamin to help soldiers delay the need for sleep. Such heroism gained him a reputation in the academic world. Following victory, he was given the chance to pursue doctoral studies at the university in Leningrad, where he set up a Laboratory for Physiological Cybernetics and was appointed professor of bionics. Now, his attention turned to the radiant energy of the human body and mind, which, he discovered, could be measured with a device called the Aurathron (figure 5.10).

For practical purposes, the key aspect of the auratic field is that it does away with the need for physical contact when gathering data from internal organs. Without mechanical operations or intrusion—just from scanning the space around the body—Gulyaev obtained electroauragrams. This data, he contended, offered empirical proof that living organisms are not bounded by their bodily frames. Instead, vital functions traverse vast distances at the speed of light. Electrical auras represent bioinformation, signals to be exploited in the field of sensory and neural bionics.

**5.11**
Information signal "psikhon"

## 5.3 Psikhon—the Biomagnetic Medium

The immaterial model of communication Khlebnikov had presented, in literature, as a figuration of the technological imaginary—an arrangement that was subsequently uncoupled from scientific dispositives under Stalin—turned into a telepathic switchboard operating in the Soviet ether at the height of the Cold War. Gulyaev, at his laboratory for physiological cybernetics, performed experiments offering positive proof for the biomagnetic medium: the information signal *psikhon* (figure 5.11, detail of Fig. 1.1).

The properties of *psikhon* enumerated in Gulyaev's diagram (figire 1.1) clearly derive from material transmission technologies. Control over media and being controlled by media are linked in a feedback system; knowledge is to be transferred mentally. The stars in lieu of brains, which point to the political dimension factoring into the equation, underwrite censorship and control. *Psikhon* stands for something that serves to take in and instrumentalize the population—an animate "agent of infection" for influencing, controlling, and steering the psyche along cybernetic lines. This picture makes it clear how both scientific insight and aesthetic practice belonged to a political-ideological program founded on the premise that mental events could directly produce real-world effects.

(83)

The flexible "mechanism" at work corresponded to the fraught mode of civil engineering that shaped the Cold War, when the art of surveillance and steering threatened to explode as soon as control mechanisms no longer agreed with their political, economic, and military machinery. Fittingly, the political-medical aspect of *psikhon*, which Khlebnikov envisioned and Gulyaev thought he could measure by means of his Aurathron, reached its apogee when the Soviet Union was in the course of collapsing and the masses had to be "recharged with healing forces."

**5.12**
Anatoly Mikhailovich Kashpirovsky,
live on television

## 5.4    1989: Media-Contagion

The year that began with George Bush taking office as the forty-first president of the United States turned out to be one of the most tumultuous in the twentieth century; ultimately, it led to the end of the Cold War.[15] Millions of people huddled in front of their television sets to witness political upheaval in the Eastern bloc: the dismantling of border facilities between Hungary and Austria, inoperative checkpoints in Czechoslovakia, and the dramatic fall of the Berlin Wall on 9 November 1989. News about protests in Tiananmen Square and the ensuing massacre prompted fascination, then outrage throughout the world. In contrast, reports that Soviet troops were withdrawing from Afghanistan occasioned jubilation and cheer.

So long as the Soviet Union remained standing, another image prevailed there. Although Channel One also issued reports about earthly events, it steered the attention of viewers—in a land that still comprised fifteen republics—toward a series of transmissions that began with an initial broadcast at 8:30, 8 October 1989, immediately after the evening news. "Relax, let your thoughts wander free," said Anatoly Mikhailovich Kashpirovsky to the citizenry of the entire Soviet Union (figure 5.12).[16] Prior to this date, the speaker—a licensed physician—had worked as a clinical psychotherapist for two and a half decades. For two years, he had also provided his services to the national weightlifting team. After the Olympic Games in Seoul (1988)—dominated by the Soviet Union, which carried away 132 medals (including six gold medals in weightlifting)—Kashpirovsky's psychic tunings achieved popularity far beyond the realm of sports. Now, the adjustments he had made in the psychophysical apparatus of athletes would calm a land beset by turbulence and heal the body politic by setting viewers' minds to the state's new goals.

Evidently, policies of reform—efforts to restructure the system with slogans of glasnost and perestroika—had failed to convince the public. It seemed the population was slowly waking up from a kind of trance and starting to realize what had been happening for the last few years; in the process, it came to recognize the gulf in the standard of living that gaped between the Soviet Union and the West, whose economic power controlled the world. Now, at this perilous juncture, a new force—one that was not manifestly political, but rather "therapeutic"—would restore the former state and rally the individual forces constituting the collective.

## 5.5 The Setting for Healing

The evening following the first of Kashpirovsky's six transmissions, 70,000 people demonstrated in Leipzig chanting, "We are the people!" This was the first protest against the GDR regime on a mass scale since 1953. But as the event was brewing, millions of Soviet citizens sat watching television as a supposed miracle worker readied them to face the future:

> You can leave your eyes open for a while. Have a look at your surroundings. There should be no pointed objects, and no fire. Your posture should be stable. If anyone is seriously ill—for example, suffering from epilepsy—please do not participate in our séance; simply turn off the television.

**5.13**
Still from the broadcast, first sequence

The first sequence of the twenty-minute session showed Kash-pirovsky speaking in close-up (figure 5.13). Then, the image shifted to the site where events were taking place. Kashpirovsky stood at the podium of a cinema, broadcasting his therapeutic message to entranced audience members as lilting musical airs filled the room (figure 5.14).

**5.14**
Still from the broadcast, second sequence

**5.15**
Still from the broadcast, third sequence

The iconography of the third, and final, sequence brought out the séance's political dimensions. As the camera panned across the movie theater, the mesmerist's face faded into, and over, the image (figure 5.15). The effect was grotesque yet sublime: like Leviathan, the sea monster of religious philosophy embodying state power in Thomas Hobbes's political theory.[17] Nor were effects meant to be limited to the sphere of images. By imbibing "charged" water, spectators were told they would feel the effects of the therapeutic session until the next transmission. At the beginning of the broadcast, Kashpirovsky instructed people seated at home to have a vessel filled with water at the ready; drinking it would revive the unconscious of the land, which had fallen ill, and instill the aims of communism in the popular mind yet again.

Kashpirovsky's third transmission took place on 5 November; the fourth occurred two weeks later. Between the two séances—on Friday, 10 November—Channel One reported that the Berlin Wall had been opened the previous evening. Both before and after this momentous event, audiences throughout the Soviet Union tuned in to be rewired and receive "star messages" so their faltering image of the world might come into focus again.

In the interference of a network where the symbolic, systematic, and rhetorical organization of data and information came into play, the specifically Soviet mode of historiographic representation was exposed: a coupling of material and immaterial practices (figure 1.1) meant to fuel the corporate enterprises of the Cold War. At the end of this confrontation of two systems, East and West—which each coordinated the assembled opinions [Meinungsbilder] of the public to strategic ends—people on the Soviet side sat engrossed by a televisual spiritist session: a miracle cure, technologically induced intoxication to dam up doubts and banish secret thoughts into the "iron cage of the unconscious." Kashpirovsky's transmission represents the last effort of Soviet power to initiate the citizenry into the mysteries of the communist apparatus that was in the course of disappearing.

**6.1**
Pavel Pepperstein's *Hypnosis of Signs*

# 6

## Aftermath: Under the Spell of Hypnosis

"We all live under the spell of signs," the contemporary Russian artist Pavel Pepperstein observes in a text written to accompany his film *Hypnosis*:

> The energy that enables the sign to hold us in a hypnotized state stems from two sources. The first source is the functioning of the sign in the sphere of its perception (that is, the "presence" of the sign). The second source is the sign's history, its mysterious past, which includes the highly enigmatic moment of how something is transformed into a sign that was not a sign before.[1]

Pepperstein seeks to expand this "highly enigmatic moment" by pursuing a psychoanalytic question: "What distinguishes the phallus from the penis?"[2] The thirty-minute film focuses on the transition of the one into the other. The conventional answer—that the phallus is a sign, whereas the penis is a body part that does not belong to any symbolic order—undergoes multilateral visual reorganization. In order to achieve the status of "universal signification," the "mere" penis—which, by itself, does not signify—must pass into the state of erection.[3] This movement, according to Pepperstein, represents one of the most mesmerizing omissions of our culture,

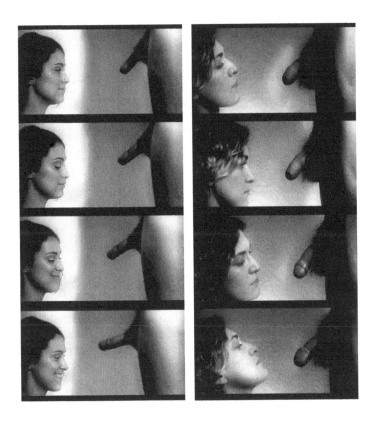

**6.2 and 6.3**
*Hypnosis*, first and second sequence

a kind of blind spot. His film offers a concrete view of this unviewable abstraction [*Anschauung dieses Unanschaulichen*]—a process that is invisible in symbolic terms.[4]

There can be no doubt that Pepperstein enlists the Freudian concept of penis envy to lay the blind spot bare (as it were). In *Hypnosis*, young women look at and animate the male member. The six sequences all show the same arrangement (figures 6.1 and 6.2): as six female faces stare fixedly, six penises move in swollen confusion and try to achieve the standing position. It is as if the women were trying to hypnotize their counterparts into the phallic state so that they might finally yield to the "magic of the sign." That said, Pepperstein reverses the polarity of Freud's famous—and controversial—thesis. As the film proceeds, it gives rise to questions. Is the woman or the penis the subject or the object of hypnosis? Does the female gaze hypnotize the male member, or is the woman herself hypnotized by the process of erection? Here, Pepperstein contends, one can sense the twitches attending a non-sign as it undergoes transformation into a sign:

> The rhythm of these convulsions resembles a dance and broadcasts certain information, so that a visualization (or the attempt at visualization) of the "secret text" of our culture emerges, in which signs and conflicts are united by an inseparable bond.[5]

In this context—and in conclusion—a brief discussion of the psychoanalyst Victor Tausk is in order. His well-known case history bears on the questions raised by Pepperstein.

In 1919, a former philosophy student, Natalija A., sought the services of Freud's loyal pupil. The young woman told him that for six and a half years, a strange, electrical device—forbidden by the police—had been manipulating her, especially her thoughts. Tausk remarked that the apparatus, as described, had "the form of a human body, namely the patient's"; it "consist[s] of electric batteries, which are supposed to represent the internal organs."[6]

Yet the patient could not say anything about how the devilish contraption functioned. She spoke only of waves, rays, and electrical currents. Tausk took note: "she vaguely thinks that it is by means of telepathy." For all that, Natalija A. was certain that a rejected suitor—a university professor—was the one operating the machine. With this device, he managed to generate or remove thoughts, feelings, and even bodily sensations at will. Tausk: "At an earlier stage, sexual sensations were produced in her through manipulation of the genitalia of the machine; but now the machine no longer possesses any genitalia."[7]

**6.4**
Pavel Pepperstein,
"Hypnosis Drawing 10"

The sessions did not last for long. After just three meetings, Natalija A. was convinced that Tausk stood under the influence of the apparatus, too—like many other people with whom she stood in contact. The psychoanalyst had come to symbolize the enemy. Because she no longer trusted him, she broke off treatment. One year later, Tausk published an account in the *Internationale Zeitschrift für ärztliche Psychoanalyse*. Then, in 1920, he committed suicide.

**6.5**
Pavel Pepperstein,
"Hypnosis Drawing 11"

**6.6**
Pavel Pepperstein,
"Hypnosis Drawing 8"

Tausk's article, "On the Origin of the 'Influencing Machine' in Schizophrenia," presents a clinical picture of the belief that one's thoughts and actions are being steered by some kind of device. Although other cases receive mention, Natalija A. occupies center stage. After reviewing parallels and analytic conclusions in the Freudian corpus, Tausk comes back to his point of departure to explain "the basis for belief that the ... influencing machine can be a projection of the patient's body."[8] He concludes:

The construction of the influencing apparatus in the form of a machine … represents a projection of the entire body, now wholly a genital. The fact that the machine in dreams is nothing but a representation of the genital raised to primacy in no way contradicts the possibility that it is in schizophrenia a symbol of the entire body conceived as a genital. … Here the genital is merely a symbol of a sexuality older than symbolism [*Symbolik*] and any means of expression suited to human interaction; therefore it can use no contemporary expression for communication.[9]

**6.7**
Pavel Pepperstein,
"Hypnosis Drawing 9"

**6.8**
Pavel Pepperstein,
"Hypnosis Drawing 4"

Although he does not quite say it, what Tausk means is that the unconscious—to which Natalija A.'s case has granted him access—amounts to a machine that makes symbols. In the Radio Age, electromagnetic waves guide its operations. Inasmuch as the influencing machine can change its shape—that is, inasmuch as it possesses prosthetic power—and is supposed to have affected all the patient's acquaintances, too, its technological functionality includes a collective unconscious, which can be represented only as "the entire body conceived as a genital."

Most of the time, the male organ in *Hypnosis* occupies an intermediate state. By itself, it calls to mind associations from the realm of natural science, archaic mythology, or science fiction. As a point of contrast, each of the film's female faces evokes the cultural inheritance and beauty-ideal of a distinct epoch (antiquity, Renaissance, avant-garde, pop art, and so on). Notwithstanding the ludic, poststructuralist perspective the director seeks to convey, the film does not indulge in anything frivolous or salacious. As the penises perform unpredictable movements, the women's smiling, enthusiastic, and curious faces do not exhibit arousal or lust. Nothing pornographic occurs. Instead, the film displays a markedly lyrical and absurd humorousness; it achieves this artful displacement by means of prolongation and delay.

This aspect is underscored by watercolor images Pepperstein made in conjunction with the film. The pictures represent a pattern of design, as if the sequence of scenes had followed their specifications (figures 6.4–6.10). Here, however, the attentive gazes from women's faces (also presented in profile) are already directed at integral symbols—for instance, a red star, a crescent moon, a swastika, or a black square. The unmarked quality of the signs—their indeterminate symbolic status—has vanished: they stand as signifiers of different cultures, ideologies, and traditions. As such, the counterparts to the symbols have fallen under the sway of the influencing machinery that transforms humankind as a whole.[10]

**6.9**
Pavel Pepperstein,
"Hypnosis Drawing 7"

**6.10**
Pavel Pepperstein,
"Hypnosis Drawing 6"

The drawings add a striking element to the questions the film poses about subject and object. Representing so many snapshots, as it were, of emergent thought-formations, they illustrate that hypnosis by means of established signs makes little (active) sense; one simply stands under their influence and "takes it." In contrast to these images, where everything is defined, it seems that the film presents an organ belonging to a nonhuman entity—an Other, a kind of sensory homunculus in "dialogue" with an icon of culture (that is, the beautiful female face).

Put in more pointed terms: one can draw a parallel between the genital striving to become a phallus and Soviet power, which—like every force destined for great influence—emerges from neural prostheses. The gentle stimulation on display equals the material foundations of telepathy (figure 1.1) we have described, which aim to hold sway over the masses, control them, and install "star thoughts" [Stern-Gedanken] that, once up and running, no longer require direct guidance.

At the same time, it is evident in Pepperstein's Gesamtkunst-werk (comprising fixed pictorial representation and moving images) that the women's gazes seek, more than anything else, to keep things at an intermediary stage—at the threshold of erection. Inasmuch as the process of solidifying-into-a-sign is consciously obstructed, the hypnotic power of the influencing machine does not prevail. In this light, neural prostheses aimed at domestic or foreign affairs—and all attendant practices—come to resemble a penis that cannot turn into a phallus. No strategy for achieving this end—whether experimental, social, or political—can be detached from underlying methods of observation, fixation, representation, and interpretation.

In each instance, the exchange assumes different form. It may proceed as flirtation, temptation, or interrogation. It can also occur through meditation, investigation, or simply conversation. Whatever the case, the parties involved remain floating in the empty, expanding sphere of hypnosis.

**Notes**

## Chapter 1

1. An almost unsurveyable number of publications on neuropros-
   thetics has appeared in recent years. Without any judgment
   being implied, one might note contributions by John. K. Chapin
   and Miguel A. L. Nicolelis, "Neuroprothesen—Roboter steuern
   von Geistes Hand," *Spektrum der Wissenschaft* 1 (2003),
   as well as Frank W. Ohl and Henning Scheich, "Neuroprothetik:
   Hightech im Gehirn," *Gehirn & Geist* 10 (2006): 64–67.

2. Sigmund Freud, *Civilization and Its Discontents,* trans. James
   Strachey (New York: Norton, 2005), 76.

3. The foundational study was authored by Freud's nephew:
   Edward Bernays, *Propaganda* (New York: H. Liverwright, 1928).
   See also Stephan J. Chorover, *From Genesis to Genocide:
   The Meaning of Human Nature and the Power of Behavior
   Control* (Cambridge, Mass.: MIT Press, 1980); Edward Hunter,
   *Brainwashing: The Story of the Men Who Defied It* (New York:
   Pyramid, 1956); David Seed, *Brainwashing: The Fictions of Mind
   Control—A Study of Novels and Films Since World War II,
   A Study in Cold War Demonology* (Kent: Kent State University
   Press, 2004); and Dominic Streatfeild, *Brainwash: The Secret
   History of Mind Control* (New York: Macmillan, 2008). Cf.
   Karl Schlögel, *Moscow 1937* (Cambridge: Polity, 2014); Jörg
   Baberowski, *Der Rote Terror: Die Geschichte des Stalinismus*
   (Frankfurt a. M.: Fischer, 2007); Slavoj Žižek, *In Defense of
   Lost Causes* (London: Verso, 2008).

4. Gilles Deleuze and Felix Guattari, *Anti-Oedipus: Capitalism and
   Schizophrenia*, trans. Robert Hurley, Mark Seem, and Helen R.
   Lane (Minneapolis: University of Minnesota Press, 1983), 240–

247. For a psychoanalytic interpretation of the mechanics of influence dating from 1919, see Victor Tausk, "On the Origin of the 'Influencing Machine' in Schizophrenia," in *Essential Papers on Psychosis*, ed. Peter Buckley, M.D. (New York: New York University Press, 1988), 49–77.

5. Leon Trotsky, *Literature and Revolution,* trans. Rose Strunsky (Chicago: Haymarket, 2005), 206.

6. Ibid., 207. Cf. Kendall E. Bailes, *Technology and Society under Lenin and Stalin: Origins of the Soviet Technical Intelligentsia 1917–1941* (Princeton: Princeton University Press, 1978).

7. My thanks to Stefan Rieger for the interpretation of these two figures and the saw. See Stefan Rieger, "'Bipersonalität' Menschenversuche an den Rändern des Sozialen," in *Kulturge-schichte des Menschenversuchs im 20. Jahrhundert,* ed. Birgit Griesecke, Marcus Krause, and Nicolas Pethes̊ (Frankfurt a. M.: Suhrkamp, 2009), 181–198.

8. I would like to thank Viktor Mazin and Joulia Strauss, who made me aware of Gulyaev's papers and entrusted them to me. The wealth of material has yielded several essays, which inform the book at hand; this study is meant as a further investigation of Gulyaev's legacy and, as such, represents a project to be developed more fully in the future. Cf. Wladimir Velminski, "Triumph des Symbolischen: Fernsehgraphische Leidenschaften in der frühen Sowjetunion," in *Bildtelegraphie: Eine Medienge-schichte in Patenten*, ed. Albert Kümmel-Schnur and Christian Kassung (Bielefeld: transcript, 2012), 235–254; "Werkbankvi-sionen: Die Einstellung der Arbeitswissenschaft aus der Poesie des Hammer-schlags," in *Ultravisionen: Zum Wissenschafts-verständnis der künstlerischen Avantgarden, 1910–1930*, ed. Sabine Flach and Margarete Vöhringer (Munich: Fink, 2009), 129–146; "Die Herr-schaft des Schweigens: Von medialen Resonanzfeldern und ihren Auswirkungen," *Resonanz: Potentiale einer akustischen Figur*, ed. Karsten Lichau, Viktoria Tkaczyk, and Rebecca Wolf (Munich: Fink, 2009), 177–192; "Krieg der Gedanken: Experimentelle Lesesysteme im Dienste der Telepathie," in *Laien, Lektüren, Laboratorien: Künste und Wissenschaften in Russland 1860–1960*, ed. Mathias Schwartz,

Wladimir Velminski, and Torben Philipp (Frankfurt a. M.: Peter Lang, 2008), 393–413; "Denken in Modellen: Zur Lösung des Königsberger Brücken-problems," in *Mathesis & Graphé: Leonhard Euler und die Entfaltung der Wissenssysteme*, ed. Horst Bredekamp and Wladimir Velminski (Berlin: Akademie, 2008).

9. Cf. Philipp Sarasin, "Infizierte Körper, kontaminierte Sprachen: Metaphern als Gegenstand der Wissenschaftsgeschichte," in *Geschichtswissenschaft und Diskursanalyse* (Frankfurt a. M.: Suhrkamp, 2003), 191–230, as well as Schwartz, Velminski, and Philipp, *Laien, Lektüren, Laboratorien*. (101)

10. Cf. Michael Hagner, ed., *Ansichten der Wissenschaftsge-schichte* (Frankfurt a. M.: Fischer, 2001).

## Chapter 2

1. Aleksei Kapitonovich Gastev, *Poeziya rabochego udara* (Moscow: Khudozhestvennaya literatura, 1971), 273.

2. Ibid., 276.

3. According to Malevich, a new, "white" humanity would emerge; cf. Boris Groys, *The Total Art of Stalinism: Avant-Garde, Aesthetic Dictatorship, and Beyond,* trans. Charles Rougle (London: Verso, 2011), 15–19.

4. Thus, the director Vsevolod Meyerhold recognized the comple-mentary aspects of scientific-rational and poetic-ludic bearings; he had Gastev and his colleagues hold lectures at the Academy of Theater Arts in Moscow. When Meyerhold called his newly developed method of acting "biomechanical," he was directly referring to the research Nikolai Bernshtein conducted in the Biomechanics Laboratory at Gastev's institute. Cf. Nikolai Bernshtein, *Ocherki po fiziologii dvizhenii i fiziologii aktivnosti* (Moscow: Meditsina, 1966), and, for commentary, Irina Sirotkina, "Istoriya tsentral'nogo instituta truda: voploshchenie utopii?," *Voprosy istorii estestvoznaniya i tekhniki* 2 (1991): 67–72.

5. This program emerged from the "Left Front of the Arts" (LEF) in the Soviet Union, an artistic association that was founded 1922

in Moscow; members included Mayakovsky, Brik, Rodchenko, Tretyakov, and many others. It existed until the end of the decade. Cf. Nikolai F. Chuzhak, ed., *Literatura fakta: Perviy sbornik materialov rabotnikov LEFa* (Moscow: Federatsiya, 1929).

6. Aleksei Kapitonovich Gastev, *Kak nado rabotat': Prakticheskoe vvedenie v nauku organizatsii truda* (Moscow: Ekonomika, 1966), 17–19 and *Trudovye ustanovki* (Moscow: Ekonomika, 1973), 54–63. For discussion, see Kendall E. Bailes, "Alexei Gastev and the Soviet Controversy over Taylorism, 1918–24," *Soviet Studies* 29, no. 7 (1977): 373–394.

7. Cf. Leon Trotsky, *Literature and Revolution*, 154–186.

8. Ibid., 48. Cf., also, Sirotkina, "Istorija tsentral'nogo instituta truda" and Rolf Hellebust, "Aleksei Gastev and the Metallization of the Revolutionary Body," *Slavic Review* 56, no. 3 (1997): 500–518.

9. Cf. Gastev, *Kak nado rabotat'*, 5.

10. Aleksei Kapitonovich Gastev, *Poeziya rabochego udara,* 19. (Translation follows *A Treasury of Russian Verse,* ed. Avrahm Yarmolinsky [New York: Macmillan Company, 1949], 252.)

11. Already in 1855, under the direction of Carl von Siemens, this branch was established as a key support for Berlin operations. Cf. Jürgen Kocka, *Unternehmensverwaltung und Angestelltenschaft am Beispiel Siemens 1847–1914* (Stuttgart: Klett, 1969), 117–127.

12. Gastev, *Poeziya rabochego udara,* 12.

13. See Hanno Möbius, *Montage und Collage: Literatur, bildende Künste, Film, Fotografie, Musik, Theater bis 1933* (Munich: Fink, 2000), 109–117.

14. Gastev, *Poeziya rabochego udara,* 12–14.

15. Margarete Vöhringer (*Avantgarde und Psychotechnik: Wissenschaft, Kunst und Technik der Wahrnehmungsexperimente in der frühen Sowjetunion* [Göttingen: Wallstein 2007]) has examined this dynamic in a brilliant study focusing on psychotechnical experiments conducted at *Vkhutemas* (the acronymic designation for "Higher Art and Technical Studios") in Moscow. She shows how Nikolai Ladovsky, who tried to determine spatial

perception objectively by means of special apparatuses and instruments, also sought to shape the experience of space in his architectural designs. See also René Fülöp-Miller, *Geist und Gesicht des Bolschewismus: Darstellung und Kritik des kulturellen Lebens in Sowjet-Russland* (Zurich: Amalthea, 1926).

16. Gastev, *Trudovye ustanovki*, 138–145. Cf. Bernshtein, *Ocherki po fiziologii dvizhenii.*

17. Gastev met with Lenin for the last time in June 1921 to discuss findings before, in August, a decree was issued (*Dekret Soveta Truda i Oborony o ZITe*). Gastev, *Kak nado rabotat'*, 7. Cf., also, Aleksei Kapitonovich Gastev, "Svidanie s Leninym," *Lenin vsegda s nami: Vospominaniya sovetskikh i zarubezhnykh pisatelei*, ed. Nina Ivanovna Krutikova (Moscow: Khudozhest-vennaya literatura, 1967), 402–404.

18. Fredrick Winslow Taylor, *Scientific Management* (New York: Harper & Brothers, 1947). Cf. Bettina Heintz, *Die Herrschaft der Regel* (Frankfurt a. M.: Campus, 1993); Angelika Ebbinghaus, *Arbeiter und Arbeitswissenschaft: Zur Entstehung der "wissenschaftlichen Betriebsführung"* (Opladen: Westdeutscher Verlag, 1984).

19. Gastev, *Kak nado rabotat'*, 29. The manifesto, *How to Work,* was initially presented at the *First All-Russian Conference on the Scientific Organization of Labor,* held 21 January in Moscow. Gastev had already devised the first rules in 1917–1918, while employed as secretary for the All-Russian Union of Metalworkers. Cf. Gastev, *Poeziya rabochego udara,* 270–272 and 297.

20. Gastev, *Trudovye ustanovki,* 62f. This same perspective also occasioned vehement criticism; opponents included Lenin's wife, Nadezhda Krupskaya, and Anatoly Lunacharsky. Cf. Nadezhda K. Krupskaya, *K voprosu o podgotovke rabochey sily* (Moscow: Gospotilizdat, 1979); Anatoly Lunacharsky, "Vospitanie novogo cheloveka," in *O vospitanii i obrazovanii* (Moscow: Pedagogika, 1968). For commentary, see Bailes, "Alexei Gastev and the Soviet Controversy over Taylorism."

21. Gastev, *Trudovye ustanovki,* 153–156.

22. Ibid., 135.

23. Ibid., 136. Cf. Michael Hagemeister, "Unser Körper muss unser Werk sein: Beherrschung der Natur und Überwindung des Todes in russischen Projekten des frühen 20. Jahrhunderts," in *Die Neue Menschheit: Biopolitische Utopien in Russland zu Beginn des 20. Jahrhunderts,* ed. Michael Hagemeister and Boris Groys (Frankfurt a. M.: Suhrkamp, 2005), 19–67.

24. Gastev, *Kak nado rabotat',* 284.

25. Cf. Kurt Johannson, *Aleksej Gastev: Proletarian Bard of the Machine Age* (Stockholm: Almqvist & Wiksell, 1983), 24.

26. Gastev, *Trudovye ustanovki,* 131. Significantly, Gastev also adopts the German word (*Einstellung*) and points to corresponding terms in French (*montage, réparation, préparation, installation*); the sole Russian word he uses is *ustanovka.*

27. Gastev, *Poeziya rabochego udara,* 264.

28. Ibid., 265.

29. Gastev, *Trudovye ustanovki,* 62f.

30. Gastev, *Poeziya rabochego udara,* 112.

31. Gastev, *Trudovye ustanovki,* 132.

32. Cf. Karl Krall, *Denkende Tiere* (Berlin: Engelmann, 1912).

33. Gastev, *Trudovye ustanovki,* 132. Cf. Slava Gerovitch, "Love-Hate for Man-Machine Metaphors in Soviet Physiology: From Pavlov to 'Physiological Cybernetics,'" *Science in Context* 15, no. 2 (2002): 339–374.

34. Gastev, *Trudovye ustanovki,* 132.

35. Ibid. 133. Cf. Torsten Rüting, *Pavlov und der neue Mensch: Diskurse über Disziplinierung in Sowjetrussland* (Munich: Oldenbourg, 2002), as well as Daniel Philip Todes, *Pavlov's Physiology Factory: Experiment, Interpretation, Laboratory Enterprise* (Baltimore: Johns Hopkins University Press, 2002).

36. Thus the encyclopedia entry "Ustanovka" in *Orga-Kalendar' CIT*, 45, quoted by Barbara Wurm, "Gastevs Medien: Das 'Foto-Kino-Labor' des CIT" (Schwartz, Velminski, and Philipp, *Laien, Lektüren, Laboratorien,* 247–290; here: p. 363).

37. Gastev, *Trudovye ustanovki,* 121.

38. Ibid., 185 (footnote).

39. Ibid.

40. Ibid., 186–188.

41. This point represents a kind of internal short circuit within Gastev's program as a whole—which, in turn, is symptomatic of a short circuit within the avant-garde program as a whole. Cf. Stefan Rieger, "Mediale Schnittstellen: Ausdruckshand und Arbeitshand," in *Mediale Anatomien: Menschenbilder als Medienprojektionen*, ed. Anette Keck and Nicolas Pethes (Bielefeld: transcript, 2001), 235–250; *Die Individualität der Medien: Eine Geschichte der Wissenschaften vom Menschen* (Frankfurt a. M.: Suhrkamp, 2001); *Die Ästhetik des Menschen: Über das Technische in Leben und Kunst,* Frankfurt a. M. 2002); and *Kybernetische Anthropologie: Eine Geschichte der Virtualität* (Frankfurt a. M.: Suhrkamp, 2003). (105)

42. Gastev, *Trudovye ustanovki,* 121. Cf. Slava Gerovitch, *From Newspeak to Cyberspeak: A History of Soviet Cybernetics* (Cambridge, Mass.: MIT Press, 2002).

43. Gastev, *Trudovye ustanovki,* 111–123. Cf. Wurm, "Gastevs Medien," as well as Siegfried Zielinski, *Deep Time of the Media: Toward an Archaeology of Hearing and Seeing by Technical Means,* trans. Gloria Custance (Cambridge, Mass.: MIT Press, 2006), 227–254.

44. Gastev, *Trudovye ustanovki,* 138.

45. Thus, Viktor Shklovsy's classic 1916 essay, "Art as Technique," introduced the concept of *defamiliarization (ostranenie)* to literary theory. One of his premises is that everyday language is automatized: speakers recognize words without sensing them; inasmuch as literature involves the process of reading, the sensory dimension is revitalized. Likewise, Roman Jakobson defined poetic language in terms of "utterance focused on expression." Cf. Slava Gerovitch, "Roman Jakobson und die Kybernetisierung der Linguistik in der Sowjetunion," in *Die Transformationen des Humanen: Beiträge zur Kulturgeschichte der Kybernetik,* ed.

Michael Hagner and Erik Hörl (Frankfurt a. M.: Suhrkamp, 2008), 229–274.

46. Cf. Rieger, *Die Individualität der Medien, Die Ästhetik des Menschen*, and *Kybernetische Anthropologie*.

47. Gastev, *Kak nado rabotat'*, 93f.

48. Ibid.

49. Ibid. Cf. Stefan Rieger, "Innovationsdruck. Zur Rhetorik der Erfindung," in *Rhetorik als kulturelle Praxis*, ed. Renate Lachmann, Riccardo Nicolosi, and Susanne Strätling (Munich: Fink, 2006), 273–290, and Gerovitch, *From Newspeak to Cyberspeak*.

50. Gastev, *Poeziya rabochego udara*, 215.

51. Quoted in A. P. Kolesnikov, *Istoriya izobretatel'stva i patentogo dela* (Moscow: Informatsionno-izdatel'skiy tsentr Rospatenta, 2005), 62. Cf. Kendall E. Bailes, *Science and Russian Culture in an Age of Revolutions: V. I. Vernadsky and His Scientific School 1863–1945* (Bloomington: Indiana University Press, 1990).

52. The parallel to the new program of constructivism (1919) is striking, which declared the machine the model to follow, as well as the ensuing agenda of *LEF* stressing the art of production and social works. For Groys, the avant-garde's failure is rooted in this "rush for political power" (*The Total Art of Stalinism*, 20).

53. *Patent na izobretenie*. ROSPATENT-Archiv, Klass. 87b, 6. Nr. 3275, v 09 V 19/24, pp. 171–172.

54. According to Spieker, the avant-garde viewed the human being, from the inception, merely as the phantasmatic counterpart of the ill, or defective, individual. Thus, he describes how the seemingly "hale" human body, both for Dziga Vertov and Vladimir Tatlin, was in fact "unfinished, incapable of movement, and basically monstrous"—which is why the projects for *Kino-Eye* and *Letatlin* came about. See Sven Spieker, "Orthopädie und Avantgarde: Dziga Vertovs Filmauge aus prothetischer Sicht (Der Mann mit der Kamera)," in *Apparatur und Rhapsodie: Zu den Filmen des Dziga Vertov*, ed. Natasha Drubek-Meyer and Jurij Murasov (Frankfurt a. M.: Peter Lang, 2000), 147–169; here: p. 150.

55. Gastev, *Trudovye ustanovki,* 71. It is striking that Gastev does not speak of prostheses anywhere in his theoretical writings on the organization of labor, thereby excluding the possibility of human imperfection. Even in his *Education of the Handicapped* (*Obuchenie defektivnykh*), he indicates that prosthetic remedies are insufficient; instead, focus should fall on enhancing mental settings. Cf. ibid., 69–73.

56. Gastev, *Kak nado rabotat',* 14.

57. Michel Foucault, *"Society Must Be Defended": Lectures at the Collège de France 1975–1976,* trans. David Macey (New York: Picador, 2003), 253.

58. Gastev, *Kak nado rabotat',* 144.

59. Gastev, *Poeziya rabochego udara,* 276.

60. Ibid.

## Chapter 3

1. Mikhail Buyanov, *Lenin, Stalin i psichiatriya* (Moscow: Rossiiskoe obshchestvo medikov-literatorov, 1993), 70. Cf. Alexander Etkind, *Eros des Unmöglichen: Die Geschichte der Psychoanalyse in Rußland* (Leipzig: Kiepenheuer, 1996).

2. Ibid., 69.

3. After removal of the brain, Bekhterev's corpse was immediately cremated and his ashes given to his wife. Postmortem analysis, performed by Aleksei Abrikosov, determined that Bekhterev had died of food poisoning. Cf. Buyanov, *Lenin, Stalin i psichiatriya,* 69–71.

4. Initial efforts to transfer thoughts were conducted around 1874 by the American Frank Brown, who organized public events featuring "mind-reading." A few years later in London, Irving Bishop—who gained notoriety in his day—surpassed his predecessor dramatically. In 1882, the Society for Psychical Research was founded here; its aim was to collect and assess all known reports of such phenomena. See Richard Baerwald, *Gedanken lesen und Hellsehen* (Berlin: Ullstein, 1933). Five years later, the

results of investigations were evaluated and the term *telepathy*—
which was meant to separate the transference of thoughts from
mental suggestion—came into circulation. Interest in Russia
had started in 1884, when Bishop performed his experiments in
St. Petersburg. In this context, the following works deserve
mention: Ivan Sikorskiy's article "O chtenii myslei" ("On the
Reading of Thoughts," 1884), Ivan Tarkhanov's study *Gipnotism
i chtenie mislei* (*Hypnosis and Thought-Reading*, 1886), Jakov
Zhuk's study on invisible communication between physical
bodies (*Vzaimnaya svyaz' mezhdu organizmami*, 1902), Naum
Kotik's books *Chtenie myslei i N-luchi* (*Thought-Reading
and N-Rays*, 1904) and *Die Emanation der psychophysischen
Energie: Eine experimentelle Untersuchung über die unmittelbare Gedankenübertragung im Zusammenhang mit
der Frage über die Radioaktivität des Gehirns* (*Emanation of
Psychophysical Energy: An Experimental Investigation
of Im-mediate Thought-transfer in Relation to the Question
of the Brain's Radioactivity,* 1908), as well as Konstantin
Kudryavtsev's *Emanatsiya psichofizicheskoy energii* (*Emanation of Psychophysical Energy*, 1908). See, in general, Eduard
Kolchinsky and Manfred Heinemann, eds., *Za "zheleznym
zanavesom": Mify i realii sovetskoy nauki* (St. Petersburg:
Dmitriy Bulanin, 2002).

5. Friedrich Kittler, *Discourse Networks 1800/1900,* trans. Michael
   Metteer (Stanford: Stanford University Press, 1992).

6. Vladimir Bekhterev, *Ob opytakh nad "myslennym" vozdeyst-
   viem na povedenie zhivotnykh* (Leningrad: AN SSSR, 1919), 8.

7. Cf. Naum Kotik, *Die Emanation der psychophysischen Energie:
   Eine experimentelle Untersuchung über die unmittelbare
   Gedankenübertragung im Zusammenhang mit der Frage über
   die Radioaktivität des Gehirns* (Wiesbaden: Bergmann, 1908).

8. Bekhterev, *Ob opytakh,* 10–11.

9. Vladimir Bekhterev, *Vnushenie i ego rol' v obshchestvennoy
   zhizni* (St. Petersburg: Aleteya, 2001), 24.

10. Sigmund Freud, *The Complete Letters of Sigmund Freud to
    Wilhelm Fliess, 1887–1904*, 146 (20 October 1895). For

commentary, see Mai Wegener, *Neuronen und Neurosen: Zum psychischen* Apparat *bei Freud und Lacan: Ein historisch-theoretischer Versuch zum sogennanten Entwurf einer Psychologie von 1895* (Munich: Fink, 2004).

11. Cf. Etkind, *Eros des Unmöglichen*; James L. Rice, *Freud's Russia: National Identity in the Evolution of Psychoanalysis* (New Brunswick: Transaction, 1993); Martin Miller, *Freud and the Bolsheviks: Psychoanalysis in Imperial Russia and the Soviet Union* (New Haven: Yale University Press, 1998). (109)

12. Vladimir Bekhterev, "O sozdanii panteona v SSSR," *Izvestiya* 137 (06.19.1927).

13. Bekhterev, *Ob opytakh,* 17.

14. Ibid., 20.

15. Bernard Kazhinsky, *Biologicheskaya radiosvyaz'* (Kiev: Izdatel' stvo Akademii nauk Ukrainskoy SSR, 1962), 33–35.

16. Ibid., 37.

17. Ibid., 38.

18. Ibid., 41.

19. Ibid., 51.

20. Ibid., 52.

21. Ibid., 53.

22. Ibid., 56.

23. Velimir Khlebnikov, *The King of Time: Selected Writings of the Russian Futurian,* trans. Paul Schmidt (Cambridge, Mass.: Harvard University Press, 1990), 158. Cf. John E. Bowlt and Olga Matich, eds., *Laboratory of Dreams: The Russian Avant-Garde and Cultural Experiment* (Stanford: Stanford University Press, 1990).

24. Aleksandr Belyaev, "Vlastelin mira," in *Sobranie sochineniy v vos'mi tomakh*, vol. 4. (Moscow: Molodaya Gvardiya, 1963), 179.

25. Ibid., 12.

26. Quoted in Daniel P. Todes, *Ivan Pavlov: A Russian Life in Science* (Oxford: Oxford University Press, 2014), 415. Cf., also, Vladimir

Bekhterev, "Ob'ektivnoe izuchenie lichnosti," in *Izbrannye trudy po psichologii lichnosti v dvuch tomakh,* vol. 2 (St. Petersburg: Aleteya, 1999), 23–23; Ivan Pavlov, "Lektsii o rabote bol'shikh polushariy golovnogo mozga," *Polnoe sobranie sochineniy,* vol. 4 (Moscow: Gosudarstvennoe Izdat, 1951), 22–30.

27. Belyaev, "Vlastelin mira," 13.

28. Ibid., 181.

29. Ibid., 120.

30. Ibid., 148.

31. Ibid., 156.

32. Ibid., 157.

33. Ibid.

34. Ibid.

35. Aleksandr Popov, *O telegrafirovanii bez provodov* (Moscow: Gos. Izd. fiz.-mat. Literatury, 1959), 79.

36. Cf. Aleksandr Popov, "Pribor dlya obnaruzhivaniya i registrirovaniya elektricheskikh kolebaniy," *Zhurnal Russkogo fizikokhimicheskogo obshchestva* 27, sec. 1, no. 1 (1896): 1–14.

37. Cf. Moisei Rodovskiy, *Aleksandr Popov* (Moscow: Molodaya gvardiya, 2009), 173–194.

38. Kazhinsky, *Biologicheskaya radiosvyaz',* 20.

39. See Cornelius Borck, *Hirnströme: Eine Kulturgeschichte der Elektroenzephalographie* (Göttingen: Wallstein, 2005).

40. Belyaev, "Vlastelin mira," 63.

41. Claude E. Shannon, "The Mathematical Theory of Communication," in Claude E. Shannon and Warren Weaver, *The Mathematical Theory of Communication* (Urbana: University of Illinois Press, 1998), 29–115.

42. Ibid., 32; emphasis in original.

43. Cf. Philipp v. Hilgers and Wladimir Velminski, eds., *Andrej A. Markov: Berechenbare Künste. Mathematik, Poesie, Moderne* (Berlin and Zurich: Diaphanes, 2007).

44. Belyaev, "Vlastelin mira," 193–194.

45. Jean Baudrillard, *Symbolic Exchange and Death,* trans. Iain Hamilton Grant (London: Sage, 1993), 74.

46. Cf., also, Ulrich Linse, *Geisterseher und Wunderwirker: Heilssuche im Industriezeitalter* (Frankfurt a. M.: Fischer, 1996).

47. Belyaev, "Vlastelin mira," 216.

48. Cf. Byung-Chul Han, *The Transparency Society*, trans. Erik Butler (Stanford: Stanford University Press, 2015).

49. Belyaev, "Vlastelin mira," 239.

(111)

## Chapter 4

1. Leonid Brezhnev, "Rabotnikam sovetskogo televideniya, svyazi, televizionnoy promyshlennosti," *Pravda* (29 November 1981): 1. Cf., also, James von Geldern and Richard Stites, eds., *Mass Culture in Soviet Russia: Tales, Poems, Songs, Movies, Plays, and Folkore 1917–1953* (Bloomington: Indiana University Press, 1995).

2. Brezhnev, "Rabotnikam sovetskogo televideniya," 1.

3. V. I. Arkhangel'skiy, "Sovremennoe televidenie i blizhayshie svyazi v etoy oblasti," *Govorit SSR* 14–15 (1933): 8–17.

4. The first mechanical Soviet television, the B-2, was manufactured in April 1932. Cf. V. I. Arkhangel'skiy, *Televidenie* (Moscow: Gosenergoizdat, 1936), 10–16.

5. B. Shefer, *Samodel'nyy televisor* (Moscow: Detizdat ZK VLKSM, 1937).

6. Cf. Schwartz, Velminski, and Philipp, *Laien, Lektüren, Laboratorien*, 9–36.

7. The expression refers to Walter Benjamin's well-known discussion of the cinema's connection to the ideological strategy of technological invisibility (*The Work of Art in the Age of Its Technological Reproducibility and Other Writings on Media,* ed. Michael W. Jennings, Brigid Doherty, and Thomas Y. Levin [Cambridge, Mass: Harvard University Press, 2008], 33).

8. Cf. Wladimir Velminski, *Form, Zahl, Symbol: Leonhard Eulers Strategien der Anschaulichkeit* (Berlin: Akademie, 2009), 87–96.

9. Cf. Rodovskiy, *Aleksandr Popov*, 219–233. On the influence of radio on Soviet culture, see Jurij Murasov, "Das elektrifizierte Wort: Das Radio in der sowjetischen Literatur und Kultur der 20er und 30er Jahre," in *Die Musen der Macht: Medien in der sowjetischen Kultur der 20er und 30er Jahre,* ed. Jurij Murasov and Georg Witte (Munich: Fink, 2003), 81–112. And Jurij Murasov, "Sowjetisches Ethos und radiofizierte Schrift: Radio, Literatur und die Entgrenzung des Politischen in den frühen dreißiger Jahren der sowjetischen Kultur," in *Sprachen des Politischen*, ed. Ute Frevert and Wolfgang Braungart (Göttingen: Vandenhoeck & Ruprecht, 2004), 217–245.

10. A. K. Tovmasyan, *Iz istorii televideniya i fototelegrafa* (Erevan: Izdatel'stvo AN Armjanskoy SSR, 1971), 54–58.

11. The Russian patent office (ROSPATENT) holds thirty-seven designs registered to Adamian, half of which concern phototelegraphy.

12. Cf. V. I. Arkhangel'skiy, "Peredatchik pryamogo videniya," *Radiofront* 6 (1935): 43–44.

13. Johannes Adamian, *Patentschrift N° 197443, Klasse 21a. Gruppe 50, Einrichtung zum Festhalten und zur wiederholten Wiedergabe von elektrischen Bildern und Bildfolgen: Patentiert im Deutschen Reiche vom 28. März 1907.*

14. Ibid.

15. Ibid.

16. Ibid.

17. Kerstin Bergmann and Siegfried Zielinski, "'Sehende Maschinen:' Einige Miniaturen zur Archäologie des Fernsehens," in *Televisionen,* ed. Stefan Münker and Alexander Roesler (Frankfurt a. M.: Suhrkamp, 1999), 13–38; here: p. 13.

18. Cf. Friedrich Kittler, *Optical Media*, trans. Anthony Enns (Cambridge: Polity, 2010), 207.

19. Cf. Siegfried Zielinski, *Audiovisionen—Kino und Fernsehen als Zwischenspiele in der Geschichte* (Reinbek bei Hamburg: Rowohlt, 1994).

20. Klaus Simmering, "HDTV—High Definition Television: Technische, ökonomische und programmliche Aspekte einer neuen Fernsehtechnik," *Bochumer Studien zur Publizistik und Kommunikationswissenschaft* 58 (1989): 13–14. See also Wolfgang Dillenburger, *Einführung in die neue deutsche Fernsehtechnik* (Berlin: Schiele & Schön, 1950), 130–135, and *Einführung in die Fernsehtechnik. Band 1: Grundlagen, Bildaufnahme, Übertragung, Farbfernsehsysteme* (Berlin: Schiele & Schön, 1975).

21. Johannes Adamian, *Patentschrift N° 197443, Klasse 21a. Gruppe 50.*

22. Ibid.

23. See V. A. Budryad, V. E. Volodarskaya, and A. V. Jarozkiy, *Sovetskaya radiotekhnika i elektrosvyaz' v datach* (Moscow: Svyaz', 1975), 35–44.

24. Kittler, *Optical Media,* 210; translation slightly modified.

25. Johannes Adamian, *Patentschrift N° 197443, Klasse 21a. Gruppe 50, Einrichtung zum Festhalten und zur wiederholten Wiedergabe von elektrischen Bildern und Bildfolgen: Patentiert im Deutschen Reiche vom 28. März 1907.*

26. Ibid., 1–2.

27. Ibid., 2.

28. Ibid.

29. Ibid.

30. Ibid.

31. On points of intersection and overlap between the natural sciences and occultism, see Wolfgang Hagen, *Das Radio: Zur Geschichte und Theorie des Hörfunks—Deutschland/USA* (Munich: Fink, 2005), as well as Dieter Daniels, *Kunst als Sendung: Von der Telegrafie zum* Internet (Munich: Beck, 2002), and Stefan Andriopoulos, "Okkulte und technische Television," in Andriopoulos and Dotzler, *1929,* 31–53.

32. Kazhinsky, *Biologicheskaya radiosvyaz',* 20.

33. Johannes Adamian, *Patentschrift Nr. 197183, Klasse 21a. Gruppe 50, Vorrichtung zur Umsetzung der örtlichen Schwankungen eines von dem Spiegel eines Oscillographen ausgehenden Lichtbündels in Helligkeitsschwankungen einer Geißlerschen Röhre: Patentiert im Deutschen Reiche vom 12. Juli 1907.*

34. Ibid.

35. Ibid.

36. Ibid.

37. Ibid.

38. Ibid.

39. Ibid.

40. Cf. Archangel'skiy, "Sovremennoe televidenie," 9.

41. Until Germany attacked the Soviet Union, sessions were broadcast every day from 11:00 p.m. until 5:00 a.m. Transmissions had the aspect ratio of 4:3, with twelve and a half images at thirty image lines per second. Broadcasts occurred at the frequency of 271 kHz, or, alternately, 1304 kHz. (Mikhail Gleizer, *Radio i televidenie v SSSR. 1917–1963* [Moscow: Svyaz', 1965], 35.)

42. "Soobshchenie," *Pravda* (30 April 1931): 1.

43. Birgit Schneider, "Die kunstseidenen Mädchen: Test- und Leitbilder des frühen Fernsehens," in *1929: Beiträge zur Archäologie der Medien,* ed. Stefan Andriopoulos and Bernhard J. Dotzler (Frankfurt a. M.: Suhrkamp, 2002), 54–79; here: pp. 61–70. On the example of transmissions in Berlin, Schneider examines how the "test girls" were colored and arranged so they would be recognizable despite the poor image quality.

**Chapter 5**

1. Kazhinsky, *Biologicheskaya radiosvyaz'.*

2. Khlebnikov, *The King of Time*, 158–159.

3. Cf. Velminski, "Krieg der Gedanken."

4. Gennadiy Paraklet, *Telepatiya* (Donezk: Stalker, 1998), 59–70.

5. Dietrich Beyrau, ed., *Im Dschungel der Macht: Intellektuelle Professionen unter Stalin und Hitler* (Göttingen: Vandenhoeck & Ruprecht, 2000).

6. Cf. Sven Spieker, "Stalin kak medium: O sublimazii i desublimazii media v stalinskuyu epokhu," in *Sovetskaya vlast' i media*, ed. Hans Günther and Sabine Hänsgen (St. Petersburg: Akademichesky proekt, 2006), 51–58.

7. Cf. John Brooks, *Thank You, Comrade Stalin! Soviet Public Culture from Revolution to Cold War* (Princeton: Princeton University Press, 2000).

8. Ibid.

9. Larisa Boldyreva, *Fiziki v parapsikhologii: Ocherki* (Moscow: Hatrol, 2003), 14–17.

10. Pavel Gulyaev, "Biologicheskaya svyaz' deystvuet. Antologiya tainstvennykh," *Tekhnika—molodezhi* (December 1968): 14–15. Cf. Oliver Caldwell and Loren R. Graham, *Moscow in May 1963: Education and Cybernetics. An Interchange of Soviet and American Ideas Concerning Education, Programmed Learning, Cybernetics, and Human Mind* (Washington, D.C.: U.S. Department of Health, Education, and Welfare, Office of Education, 1964).

11. Ibid., 14.

12. Cf. Velminski, "Die Herrschaft des Schweigens."

13. Vladimir Bekhterev, *Suggestion und ihre soziale Bedeutung: Rede gehalten auf der Jahresversammlung der Kaiserl. Medizin. Akademie am 18.12.1897. St. Petersburg* (Leipzig: Georgi, 1899), 3–4; cf. Ute Holl, *Kino, Trance & Kybernetik* (Berlin: Brinkmann und Bose, 2002), as well as Stefan Andriopoulos, *Besessene Körper: Hypnose, Körperschaften und die Erfindung des Kinos* (Munich: Fink, 2000).

14. Cf. Vladimir Bekhterev, *Ob opytakh*, as well as Velminski, "Die Herrschaft des Schweigens."

15. According to network records in West Germany (ARD and ZDF), news about events in 1989 exceeded all viewing figures to date.

16. Alexander Torin, *Die wahre Geschichte der Extrasensologie in Russland* (1997, unpublished typescript).

17. Cf. Horst Bredekamp, *Thomas Hobbes: Der Leviathan. Das Urbild des modernen Staates und seine Gegenbilder 1651–2001* (Berlin: Akademie, 2003).

## Chapter 6

1. The film's premiere occurred at the *Art and Crime Festival* in Berlin, 2003. Quotations follow the booklet to the DVD (Moscow).

2. Pepperstein, *Hypnosis*.

3. Ibid.

4. Pepperstein points out that even pornographic films claiming to "show everything" fail to pay notice to this process.

5. Pepperstein, *Hypnosis.*

6. Victor Tausk, "On the Origin of the 'Influencing Machine' in Schizophrenia," in *Essential Papers on Psychosis*, ed. Peter Buckley, M.D. (New York: New York University Press, 1988), 57. Cf. figure 17.

7. Ibid., 57.

8. Ibid., 72.

9. Ibid., 72–33; translation modified.

10. Cf. Martin Burckhardt's afterword to Victor Tausk, *Beeinflusungs-apparate: Zur Psychoanalyse der Medien* (Berlin: Semele, 2008), 68–94; here, p. 88.

"*Homo Sovieticus* analyzes the convergence between culture and experimental science, the point at which mental energy becomes art through a series of relays from the laboratory to the studio. The book's merit is to demonstrate that the Soviet 'New Man' was missing a crucial limb: the brain. Its proper conditioning by scientists, politicians, and writers is the focus of Velminski's study."

**Sven Spieker, University of California, Santa Barbara**

"Wladimir Velminski illuminates an obscure but deeply influential stratum in the history of social control. By examining a series of Soviet patents, concepts, fictions, and events he demonstrates the deep-set cultural and scientific belief that minds, souls, and ultimately entire nations might be manipulated through the combined applications of technology and telepathy."

**Mark Pilkington, author of *Mirage Men***

Printed in the United States
by Baker & Taylor Publisher Services